Introducing HEATHER
Scotland's most remarkable plant

CONTENTS

Introduction To Heather	3
Heather Burning & Regeneration	6
Heather In The Food Chain	6/7
Associated Birds, Animals, Insects etc	9
Associated Plants	11
Heather in the Dwelling Place	12
Other Uses of Heather	16
Recipes	20
Honey	22
Healing Properties	23
The Heather Garden	24
Cultivars	37
The Heather Society	48

by David Lambie

INTRODUCING HEATHER

When God first made the world, He looked at the bare and barren hillsides and thought how nice it would be to cover them with some kind of beautiful tree or flower. So he turned to the Giant Oak, the biggest and strongest of all of the trees he had made, and asked him if he would be willing to go up to the bare hills to help make them look more attractive. But the oak explained that he needed a good depth of soil in order to grow and that the hillsides would be far too rocky for him to take root.

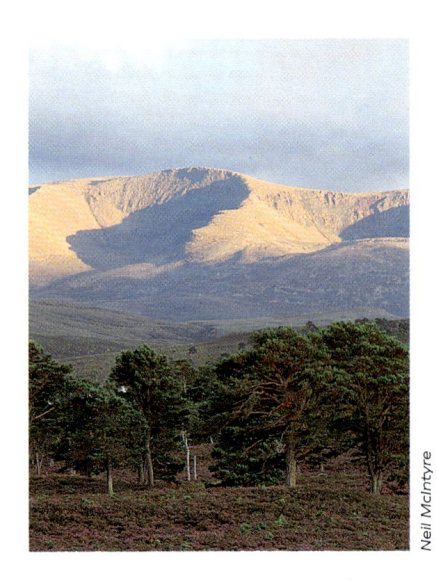

So God left the oak tree and turned to the honeysuckle with its lovely yellow flower and beautiful sweet fragrance. He asked the honeysuckle if she would care to grow on the hillsides and spread her beauty and fragrance amongst the barren slopes. But the honeysuckle explained that she needed a wall or a fence or even another plant to grow against, and for that reason, it would be quite impossible for her to grow in the hills.

So God then turned to one of the sweetest and most beautiful of all the flowers - the rose. God asked the rose if she would care to grace the rugged highlands with her splendour. But the rose explained that the wind and the rain and the cold on the hills would destroy her, and so she would not be able to grow on the hills.

Disappointed with the oak, the honeysuckle and the rose, God turned away. At length, he came across a small, low lying, green shrub with a flower of tiny petals - some purple and some white. It was a heather.

God asked the heather the same question that he'd asked the others. "Will you go and grow upon the hillsides to make them more beautiful?"

The heather thought about the poor soil, the wind and the rain - and wasn't very sure that she could do a good job. But turning to God she replied that if he wanted her to do it, she would certainly give it a try.

God was very pleased.

He was so pleased in fact that he decided to give the heather some gifts as a reward for her willingness to do as he had asked.

Firstly he gave her the strength of the oak tree - the bark of the heather is the strongest of any tree or shrub in the whole world.

Next he gave her the fragrance of the honeysuckle - a fragrance which is frequently used to gently perfume soaps and potpouris.

Finally he gave her the sweetness of the rose - so much so that heather is one of the bees favourite flowers. And to this day, heather is renowned especially for these three God given gifts.

INTRODUCING HEATHER
Scotland's most remarkable plant

Wandering the hills and moors near my home here in Strathspey, I never cease to appreciate that I live in one of the most beautiful and unspoilt areas in Europe. Surrounded by breathtaking scenery and wildlife in abundance, one of the most spectacular sights to be seen, must surely be in late August and early September when the countryside is covered in a mass of Heather blooms.

The Scottish Highlands are justifiably renowned for this famous, but often underrated plant, which grows so profusely in this part of the world. With its contribution to the scenic splendours of Scotland inestimable, the praises of Heather have been heralded in Song and Legend since time immemorial. And yet so little is known about it.

Having spent 21 of my most enjoyable years growing, researching and photographing this plant, I have been amazed at the amount of information which I have been able to gather.

From unearthing facts and stories in museums and libraries to actually savouring the delights of its honey and wine, I have gradually built up a store of information about this most remarkable plant, which I now share with you in this book. I would like to think that a small percentage of the ten million visitors from within the United Kingdom and overseas who visit Scotland annually will, through reading this book, come to appreciate even more, the plant which is synonymous with this beautiful country.

The book covers many aspects of the plant, starting with its origins and role in moorland life, and moving on to its use as a natural building material in and around the home. I have also included a section on the medicinal properties of heather.

The book concludes with some practical advice on the planning and preparation of a heather garden. From designing a layout to choosing your plants, there are plenty of useful planting suggestions and tips - things which I've picked-up and discovered over the years.

Finally, there is a comprehensive list of cultivars to look out for, along with details of their flowering times, colour, shape and foliage colour.

I have thoroughly enjoyed putting this book together and would like to thank all the kind people who have helped me along the way. From those who sent me information, stories and literature to those who typed, typed and re-typed! I sincerely hope that you enjoy reading this book, and I trust that it won't be long before you too come to regard the 'Scottish Heather' as a most remarkable plant.

SCOTLAND'S MOST REMARKABLE PLANT

Heather, the name most commonly used for this plant, is of Scottish origin, presumably derived from the Scots word HAEDDRE. Haeddre has been recorded as far back as the fourteenth century, and it is this word which seems always to have been associated with ericaceous plants.

The origination however is obscure, and the variations are many. Hader is found in Old Scottish from 1399, heddir from 1410, hathar from 1597 (although this form of the word may also be seen in place names dating back to 1094) and finally heather from 1584.

The botanical name for the Heath family is Ericaceae, which is derived from the Greek 'Ereike', meaning heather or heath. The name is generally, and more properly reserved for the most widespread of the Heath family Calluna vulgaris, (Calluna from the Greek 'Kallune' - to clean or brush as the twigs were used for making brooms and vulgaris from Latin, meaning common.)

However the plant is sometimes also referred to as Ling - derived either from the old Norse 'Lyng' or from the Anglo Saxon 'Lig' meaning fire and referring to use as a fuel.

Close up photograph of wild heather flowers (Calluna Vulgaris).

View of Strathspey from the Monadhliath mountains showing Loch Alvie and the Cairngorms.

Whatever the exact origin, one thing is certain. Heather moors cover a vast amount of Scottish countryside. With approximately 2 to 3 million acres of Heather Moors in the East and only slightly fewer in the South and West, Heather is without doubt one of Scotland's most prolific and abundant plants.

A Plant in Abundance

There are a number of reasons why heathers are so abundant with such a wide distribution. Firstly, the plant's reproductive capacity is high with seeds produced in very large numbers. Each tiny heather flower has 30 seeds, so it is quite possible for one large plant to produce up to 150,000 seeds per season. Small and light, the seeds are readily dispersed and pollinated by wind and insects, with the germination period lasting up to six months. This long period is advantageous to the heather as it means there will be reserve seedlings to take over if the first seedlings should fail.

Soils

Most heather seeds are normally shed during November and December, with germination proving most successful on soils

FIVE

INTRODUCING HEATHER

with a pH of between 4.5 and 7.5 (although it is better on soils tending to the lower end of this scale). But Heather is a very versatile plant. It readily adapts and thrives in soils which are not only acid, but also those which are poorly supplied with the mineral elements normally essential to plant growth. Heather can survive in many soil types, from those which are peaty with a high water content to those which are free draining and relatively dry.

Grazing

Despite repeated grazing, heather is relatively resistant to feeding cattle and sheep. With reserve buds readily replacing those which have been cut back by the animals, this hardy plant is really only damaged and destroyed when grazing becomes excessive.

The hardiness of heather is demonstrated by its ability to flourish despite recorded temperature extremes. At ground level, temperatures have been recorded in Strathspey as low as -32°C (-18°F) in winter and as high as 38°C (100°F) in summer.

Life expectancy of heather is approx. 40-50 years.

Heather Burning & Regeneration

One of the most common and effective ways of managing heather clad land is the age old method of burning. Heather burning or 'Muirburn', as it is called in Scotland, was, and indeed still is, necessary for a variety of different reasons.

Originally the process was used not only to prevent tall shrubs and trees taking over the moor - limiting the number of older Heather plants which were tall and woody with few young shoots. But it was also used to create open spaces of land which made it difficult for preying wolves and foxes to approach domestic herds unseen.

Now however, heather burning is primarily used to create the conditions necessary for uniform regeneration of young heather plants - encouraging high reproductivity of edible new shoots which are essential to animals and birds dependant upon heather for survival. Grouse shooting employs many people and brings in valuable revenue in sparsely populated areas of Scotland.

Burning, as you might expect, is restricted by law.

In order to cause as little disruption and damage as possible to both nests and wildlife, the period for burning in Scotland currently begins in October and lasts until 15th April. Traditionally however most burning is carried out in Spring, although some estates maintain that regeneration is better after burning in the autumn. A saying from North Uist, which has been passed down through generations, refers to the heather burning which had to be done in a bad season;

'Is fearr deathach a' fhraoich, na gaoth a reodhaidh'
(Better is the smoke of the heather than the wind of the frost.)

After burning almost all of the above ground parts are destroyed. But reserve buds at the base of the stems, which have been shielded by earth and moss, are protected from the heat and give rise to new growth the following season. This is known as vegetative regeneration, with the new growth, referred to in Gaelic as Mionaa (Meanbh) Fhraoch.

However, this is not the only regenerative process. Another, although somewhat slower way, is by seed. Germination of heather seed occurs in the gaps between the burnt plants - usually where there is a moist seed bed and fluctuating temperatures

Heather in the Food Chain
Domestic and Wild Animals

Although few animals rely on just one food source, many depend significantly on heather for survival. - Red Deer, Rabbits and Hares, to name but a few. And it is the heather's young shoots with which most of these animals supplement their diet.

SCOTLAND'S MOST REMARKABLE PLANT

Heather burning for red grouse management.

One look at a typical example of a food chain shows just how important heather is in the cycle of moorland life.

Heather shoots....eaten by caterpillars....in turn eaten by Meadow Pipit.... preyed upon by Hen Harrier.

At their best, the shoots provide important minerals such as calcium, phosphorus, nitrogen, magnesium and potassium, with their nutritional quality determined by the particular soil quality in that area.

The shoots may be grazed throughout the year. In winter, the green shoots are eaten followed by the new green shoots in May. In summer the flowers are eaten, in bud and bloom, along with the seed heads. And in autumn, even the seed capsules help to sustain wildlife on the moor.

The Red Deer browse heavily on the heather, particularly in the winter months when other food is hard to find. The Roe Deer also graze on the plant, preferring to stay in the longer heather. Even the Reindeer which were introduced into the Cairngorm area in the 1950's, have also come to rely on this popular food source. Other 'nibblers' include Mountain and Brown Hares who require the young heather for browsing, and rank heather for cover. Rabbits living on moorland also enjoy young shoots.

According to the Scots National Dictionary the Wild Cat was also referred to as the Heather Cat.

Of course domestic animals too, such as cattle and sheep, also benefit from grazing on heather - in areas which have been specially designated and managed solely for this purpose.

Cattle use the heather to supplement their winter diet of hay, turnips and manufactured foods. 'Teck' is the name used in Shetland to describe the long heather which was used as fodder. However it is perhaps, the hardiest of domestic breeds which is the main beneficiary - the Blackface sheep. For many years to the present day, sheep farmers have taken their sheep to the heather to give the resultant lamb 'that special flavour'. Indeed the Blackface Sheep Breeders Association is pushing hard to have the pure-bred 'Blackface Lamb from the Heather', officially recognised as one of Scotland's gourmet tastes, comparable with grouse, venison, whisky and salmon.

There are, however, drawbacks as the following extracts from "The Scots National Dictionary" outline:-

Heather Blindness – a disease of sheep.

Contagious opthalmia is a specific disease of sheep that is common to most sheep-raising countries, including Scotland,

Emperor Moth caterpillar feeding on heather – colour blending perfectly with its hos

SCOTLAND'S MOST REMARKABLE PLANT

Merlin (& chicks) nesting in rank heather.

Small tortoiseshell butterfly on bell heather.

where it is often referred to as "heather blindness".

Heather Cling – a disease prevalent among sheep that have been grazing too long on heather.

Heather Claw – A dog's dew claw, which is apt to catch in heather with resulting pain, and is therefore often cut off.

Heather Clout – Also in reduced form Clu.

The fettock, joint or ankle, the external and posterior part of which is protected by two horny substances, which we call heather-clouts.

A Natural Habitat for Birds

The bird most associated with heather is the Red Grouse, sometimes referred to as 'The Heather Bird'. The Gaelic term for the male bird is Coilech-Fraoch, the Heather Cock, with the female equivalent being Cearc-Fraoch, Heather Hen.

Apart from nesting time when the bird feeds primarily on insects and grubs, the Grouse will feed on young heather shoots. Using grit to grind up the hard fibrous heather in it's gizzards, the bird is able to turn much of the lignin and cellulose, (the woody part of the plant), into usable energy. So even in heavy snows when the heather can still be seen sticking up through the icy cover, there will be no lack of energy to maintain life. Indeed, this is why grouse are seldom found starving during deep snows.

Breeding tends to be more successful on more nutritious heather.

Of course, Heather is more than just a food source. Heather is also a natural habitat for the hundreds of different insects, animals and birds all of whom are inter-dependent on one another.

Providing nesting sites, shelter and protective cover from predators there are many different types of bird to be found thriving on the moor.

Red Grouse: Relies on heather not just for food but also as a protective cover from predators such as the Golden Eagle and Peregrine Falcon. In Caithness, it is common to see bunches of heather tied to the top of deer fences which warn these and other low-flying birds, to take evasive action!

Black Grouse: The best stocks of this bird are to be found on a moor with a mixture of scrub, tall heather, bogs, shorter vegetation and farmland. The hen chooses thick heather for her nest and then moves her chicks

INTRODUCING HEATHER

Hen ptarmigan and chick among purple heather and boulders (summer plumage).

to damp ground with rushes, tall heather, bog myrtle and scrub. Known as the Heather Cock, this bird consumes the Heather Beetle in great quantity.

Ptarmigan: This bird is normally found on higher slopes, but in severe winters it will come downhill in search of the young shoots.

Capercaillie: The Turkey Sized Capercaillie uses pieces of heather or pine needles to cover its eggs until its clutch is complete.

Golden Plover: This bird with its beautiful, plaintive cry nests in the short heather.

Dotterel: Like the Golden Plover and a member of the Plover family, this bird makes its nest on a carpet of prostrate heather, high in the Cairngorms.

Lapwing: Sometimes referred to as Peewit or Green Plover, this bird can be found in short heather close to the grassland of nearby crofts and farms.

Curlew: For me, Spring has arrived when the beautiful bubbling calls of this bird can be heard as it glides back from wintering at the coast to its nesting site in the long heather.

The Common Snipe: Sometimes referred to as the Heather Bleat(er), gets its name from the sound made by the male with its tail feathers in flight - mainly during courtship.

Ring Ouzel: Known as the 'Heather Blackie', this bird is very much the 'Mountain Blackbird' and makes its nest from coarse grasses, heather and mud.

The Meadow Pipit: Also known as the Heather-Cheeper.

The Heather Lintie: Also referred to as the Twite or Mountain Linnet breeds on the moors and open ground.

The Heather Peeper: Another name for the Common Sandpiper.

The Heather-Cun-Dunk: A sea bird, either the Goosander or the Northern Diver.

Golden Eagle: this magnificent bird constructs its nest, or eyrie, from twigs of heather and preys on moorland creatures such as Red Grouse and Mountain Hares. Other birds of prey which are associated with the heath include the Buzzard, Peregrine Falcon, the Merlin, Hen Harrier and Short Eared Owl.

Insects

Heathers are affected relatively little by pests. However there are certain types of insects which can be found living on the plant. These include the Heather Gall Midge (Wachtiella ericina) and the Heather Beetle (Lochmaea suturalis).

Other types of insect to live on the heather are the Sap Suckers. One such example is the Froghopper, sometimes referred to as the Spittlebug, because of the froth of bubbles with which it surrounds itself. These bubbles are created from the sap of the heather plant and are a familiar sight to many people.

The Emperor Moth is a caterpillar which feeds on the young

SCOTLAND'S MOST REMARKABLE PLANT

Left: *Short-eared owl and 2 chicks, often seen flying over heather moorland during daylight.*

Above: *Ring Ousel, the "Heather Blackie".*

shoots. As you will see from the photograph (on page 8), it is camouflaged magnificently with its green body taking on the exact colour of the heather, whilst the pimples on its body resemble the colour of the heather flower.

The adult moth has tremendous 'eyes' which help to ward off prey as it flies over the heather throughout April and May. After mating the female flies by night to lay olive-brown eggs on the heather.

The Northern Eggar Moth is another caterpillar which is also common on heather moors, feeding, once again, on the young shoots.

In Banffshire, the Dragon-Fly is better known as the Heather Bill.

Reptiles
There are a few reptiles to be found living amongst the heather. It is not unusual to see Adders basking in the sun on hummocks before slithering off to hunt for toads and frogs in nearby pools. In Caithness, frog tadpoles are called Heatherfish.

Associated Plants
On well drained areas Calluna is generally accompanied by Erica cinerea (Bell Heather), Tormentil and Common Milkwort. While on wetter soils, Bell Heather is replaced by Erica tetralix (Cross-leaved Heath) accompanied by other plants such as Bog Asphodel and Sundew.

In open pine woods, where the ground vegetation is heathy in nature with plants such as bilberry, cowberry and crowberry, and grasses like wavy haired grass and soft grass, heather is once again a prominent member of the plant community. Even in the Birch woods where the canopy is light, this versatile plant is not hindered in its growth and is surrounded with bracken and bilberry.

Carline Heather: The bell heather of Erica tetralix and Erica cinerea. Used also to describe

ELEVEN

INTRODUCING HEATHER

French Heather, Fraoch Frangach, by old Perthshire hill-folk.

Cat Heather: This is a name used to describe various types of heather including Erica cinerea, Erica tetralix, or Calluna vulgaris.

Dog Heather: This is the Aberdeenshire name given to the ling heather.

Heather in the Construction of and Thatching of the Dwelling Place

As one of the most common and readily available resources in the countryside, Heather has always played an important role in the traditional construction of buildings, particularly in areas such as the Hebridean Islands where construction was strictly determined by the availability of natural materials and their proximity to the proposed site. As a result, heather was used to build many dwelling houses, churches and farmhouses. From walls and thatching to the ropes and pegs which actually held the building together, Heather proves its versatility once again.

Thatching with Heather was carried out in areas as far apart as Shetland in the North and the Island of Arran in the South West. Buildings which were Heather Theekit - (thatched with Heather), were generally a better class than those which were thatched with straw. The old Blackhouses of Lewis for instance which were thatched with straw and constructed without a chimney or smoke hole, so as to impregnate the straw with soot for future use as a fertilizer in the fields were, despite being the possible forerunners of our present day recycling philosophy, unpopular as they had to be replaced annually.

A Heather roof however, according to J. Smith in the 'General View of Agriculture of Argyll'(1798), will last 100 years. He tells us that, "Heather roofs are more suited to Farmhouses, as they, along with our ordinary timber, can be had for a trifle, last almost as long as slates and

Heather thatch on house in Sidinish, North Uist.
Photo courtesy of National Museum, Edinburgh.

give less trouble in repairs..., It is astonishing that, in a country in which Heather abounds, these roofs are not more common. They are indeed heavier than straw roofs; but by making them a little steeper, and placing the couples a little nearer than ordinary roofs, most of the weight will be thrown on the walls, which, if made as they ought to be, of stone and lime, will not feel the burden. It makes a neat warm and durable roof."

The thatching of these roofs was done by a 'heatherer' and various methods of construction

SCOTLAND'S MOST REMARKABLE PLANT

Detail of heather thatch, Sidinish, N. Uist. Photo courtesy of National Museum, Edinburgh.

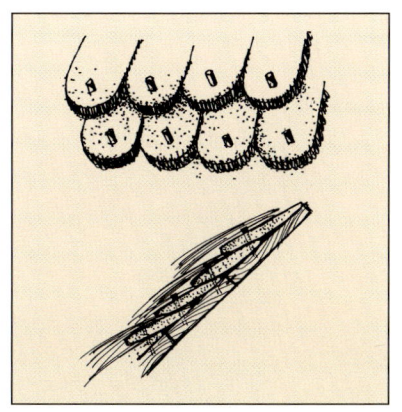

Sketch showing detail of heather divots on roof secured with heather pegs.

have been recorded. One method however, which was perhaps in most common use was to make a covering of divots (thin sods pared off by a spade made specifically for that purpose) above the wooden framework. Over that was then spread a thin coat of thatch which was then fastened down by straw or heather ropes, crossed through each other in a net-like fashion, and with stones suspended.

Heather divots would frequently be used to cover the crofters byre and stable roofs and quite often the crofter's own cottages too. But most often, they were used for covering potato and turnip pits in winter. The divots were clean and warm, and at the same time, provided better ventilation than grass sods. Using the heather divots in this way, laid on like slates, heather side down, there was no danger of the crops sweating and rotting.

The heather-thatched shielings of Glen Lyon between 1837 and 1841, were described by Duncan Campbell in his book, 'Reminices of an Octogenarian Highlander', as not much to look at on the outside, but rather substantial and roomy on the inside. Built of stone, thatched with heather and well constructed for dairy purposes, recesses, or rather cupboards, were built into the thick walls with flags for shelves on which milk vessels were placed. Planks were placed at intervals across the building from one top of the side wall to the other. And it was here that the cheeses, partly taken out of their presses, were placed to harden and become partially smoked with the reek of the peats and the remains of the heather stalks which had been burnt to the ground the year before.

In ancient times, Churches too were generally thatched with heather.

The roots of heather served as effective nails and pegs - used especially for hanging slates. But it was the small twigs of heather, trimmed and shaped with a knife which were used to peg down the heather divots in the thatch. Once pegged down, a thick fringe of heather was arranged to project under the lowest layer of divots in order to carry rainwater drips away from the roof, clear of the walls. If however the house did let in water, fesgar, a facing of straw or heather, would be fastened with boards around the outside of the house. (Fesgar is also referred to as a strengthening rim for straw or heather baskets.)

Walls were constructed of Heather-An-Dub (Heather and daub, sometimes spelt dab) which was a combination of heather with mud or clay. Built with an inner and outer skin of stone, the walls had a central core of heather divots. There were many types of resulting buildings. Basket Houses for instance on the island of Mull

INTRODUCING HEATHER

Christina MacVarish carrying home heather firewood, Bracora, Morar, Inverness-shire c.1905. Photo by courtesy of National Museums of Scotland.

were constructed by wattling together heather and branches of wood. And in Strathspey, it has been recorded that heather brushwood was used (in conjunction with wild juniper) in cavity walls of houses to act as insulation and soundproofing. And talking of insulation, it has also been used by hill people as insulation against the cold by packing it down trousers and inside jumpers!!!

ROPES AND LADDER: The Skara Brae excavation in Orkney revealed a prehistoric village dating back to around 2000BC. Amongst primitive tools and animal bones they discovered Seomain fraoich - Heather rope.

Seomain fraoich was made from long stems of the Heather plant, pulled and woven by hand, and used for a variety of purposes - from holding down thatch to securing fishing boats. It was even used in Harris to gather in the seaweed for the production of kelp which, in turn, yielded iodine essential to the manufacture of glass.

In the Helmsdale area ladders made from heather ropes, bound together and swung over the cliffs, were used to give direct access to the shore.

Inside The Home

Even to the present day, crofters and farmers rely upon heather as an abundant and efficient fuel for their fires - with the part which is most generally burnt coming from the top layer of the peat bog. Once cut, stacked and dried it was used for heating the dwelling place, cooking, drying (especially for drying corn before it went to the mill), brewing and baking. The small heather stems were also used as they were found to make excellent 'kindlers' for starting the fire. The crofters even found a use for 'Heather Birns' burnt heather - using the short, charred stalks of the plant as writing instruments!

The Skara Brae excavation in Orkney revealed evidence of beds in the form of stone boxes, lined with either heather or straw and dating as far back as 2000 BC. The original heather beds!

Obviously these beds were developed over the years with crofters gradually perfecting their construction techniques using only the longest, straightest and finest stalks of the young heath. These stalks would be pulled at their highest in bloom and fragrance, with as little root as possible. Once they had been left to dry for a few hours to evaporate any dew or accidental moisture, they would be placed together as thickly and closely as possible, with their tops arranged uppermost. Inclining a little towards the head of the bed - which was generally against a wall, the stalks would be held together at the sides and foot of the bed by logs of wood which had been cut to appropriate lengths.

Even an outdoor heather bed has its merits!

And just in case you were wondering about the support afforded by a heather bed, you might be interested to note that, according to Mr R. Oldale, Kilchrenan, Argyll, the anvils below the huge drop hammers in the Sheffield Steel forges were sited on beds of heather. These heather beds were apparently sufficiently well cushioned so as to absorb the tremendous impact delivered by the hammers and thus prevented the anvils from fracturing!

Inside the croft, heather had many practical uses - from baskets and brushes to pot scrubbers and doormats.

Heather brooms were made from the long heather stems which were gathered in spring when they were at their most pliable. These would then be bunched together and guillo-teened to form a trim broom.

Many types of besoms and brooms were made in this way, and the art of brush making became a small, rural craft industry. Types of brushes and besoms varied, depending on locality, and were sold round-abouts by Heather Jennys and Heather Jocks (the nicknames for the men and women who sold heather goods). Indeed a great trade was long practised in the making of heather goods as, until around 1860, there were few modern brushes to be found on an ordinary farm.

A Curling Heather Cow was the term used to descibe a broom made of heather twigs whilst a Heather Range(r)/Reenge - sometimes referred to in Orkney as a Heather Scratter - was the name given to a bunch of straight heather stems cut to equal lengths and bound firmly together for use as a pot scrubber or for brushing the flue of a chimney. In-fact, to this day, Chimney Cleaners in Lewis still use bunches of heather tied to to the end of ropes as they insist that it is the only way to get the job done properly.

Heather doormats were also common in the croft. Mats would be specially made for the kitchen using long, thin heather stems. They would be woven in many different patterns with the bottom side rough where the clipped ends were concealed and the upper side smooth. In Islay, these mats which were woven from young heather were known as 'peallagan'.

Baskets were made all over the highlands and had many uses round the croft. Using long heather stems, they were made either to be carried on the human back or as pack saddles which were carried by horses. They were also hung on walls for storage. The Saalt (Salt) cuddie from Shetland, is just one example of such a basket which was hung beside the fire and used to keep salt dry.

Of course there were many other different types of basket each with a specific role. The Mudag (Wool Basket) for instance, was specially made to hold wool before carding, and the Maisie - a large meshed panier of bent rope or heather, was used for carrying sheaves and peats.

In Orkney, one particular form of basket was called a 'Heather Cubby'. These baskets were made in various forms but the most common one was woven from long, fine, straight heather stalks, not rough or crinkly, and was used for carrying turnips from the shed to the byre for the cattle. Carried on the back, they were also used to bring peats inside from the stack. Another type of basket was the 'Sea Cubby' - so called because it was specially used to carry home fish. Other baskets including the Heather Caissie (from Orkney) and the Heather Wuddie/Widdie were made to hold fish, fishing line and bait. Even the lobsters of the Hebrides which were sent to London where they were in great demand, travelled packed in heather.

Examples of these and other utensils can all be seen at the Highland Folk Museum at Kingussie.

Heather "cassie" with coir yarn spliced into rim, Orkney.

Tattie basket, Kingussie.

Detail of hand broom, North Uist.

A coil of heather rope, Skye.

Uses of Heather Outdoors

Abundant and readily available, heather, with its twiggy nature and durable characteristics - even in bog conditions, was often used in the making of roads, tracks and footpaths. It was mainly laid as an intermediate layer between a base of brushwood and a surface of gravel. And it was this method, which was reportedly used, in laying tracks across Rannoch moor.

In Mediaeval times, the otherwise useless parts of a sheep's fleece, known as 'daggings', were laid and mixed with heather to form footpaths across the heath. This ancient practice is currently experiencing a revival, with daggings by the hundred-weight being airlifted by the RAF to reinforce foundations along the popular walkways of the Cairngorms.

Depending on the drainage, lighter soils can be prone to 'silt up'. To avoid this happening, the tops of old heather would be placed in the bottom of a trench to act as a field drain. Another way in which heather was useful as a 'conservation method' was in the production of stabilising banks. These banks, in conjunction with planting marram grass, would be made from long heather twigs and used to stabilise dunes - a method

SCOTLAND'S MOST REMARKABLE PLANT

Detail of doormat, Loch Eport, North Uist.

Heather broom, Kingussie.

Guillotine for heather brooms, Kingussie.

Wester Ross, heather pot scrubbers.

Photos by courtesy of The Highland Folk Museum, Kingussie.

Heather lamp being demonstrated.

employed, particularly in Holland.

Used for protecting sheep and other animals in the winter months, sectioning them off into particular areas, fences would be made from the longest, most supple stems of rank heather, intertwined between stakes and posts.

Heather branches were very often used to make walking sticks, particularly in Colonsay where the rich peaty soil on the east of the island made ideal growing conditions. Two such specimens were sent to Edinburgh University. One branch was measured at 6ft. whilst the other was a mere 4ft. in height!

Salmon Fishing

Heather was often burned as a fuel for cooking and heating. But it was also used for lighting - Salmon Lighting that is. The following extract from 'Days and Nights of Salmon Fishing in the Tweed' by William Scrope, descibes how heather was used in an age old technique for catching salmon.

"We went to the barn and tied up twae heather lights frae a bunch or twae which I had gead the miller lad to dry in the kiln ten days before. They may talk o' ruffies and birk bark baith, but gie me a good heather light, weel dried on the kiln for a throat o' the Queed."

Above: *an old lamp for heather burning. The small tank is filled with paraffin and impregnates the wick which is then ignited.*

Bunches of heather which had been thoroughly dried were placed inside a special basket-like carrying device. (See sketch page 18) Once set alight, this 'heather torch' would be held at the water's edge where the flickering flames and lights would attract the fish from the low water, (known as 'burning the water'). The salmon would approach the lights, whereupon they would be speared by a five barbed, long handled fork, called a "leister". This method of fishing was legal at the time.

SEVENTEEN

INTRODUCING HEATHER

Above: *Never missing a puff, he turns out heather rope. Twisting the strands just right is an art. The result looks rough but this South Uist islander says no other rope will outlast a well-made one of heather. It serves many purposes, notably to hold down the thatch on the little stone houses which seem to belong to the landscape.*

From Margaret Fay Shaw: Folklore and folksongs of South Uist (London, 1955).

Below: *An illustration from 'Salmon Fishing' Scrope, 1843. This shows two men spearing salmon from a boat, the 'cruisie' filled with dry heather and ignited was mounted centrally between them.*

Below right: *The cruisie illustrated, for 'burning the water' when leistering at night, was drawn from an example lent to the National Museum for exhibition some years ago by the Chambers Institute, Peebles. In Peeblesshire it is known as a 'craizie', a variant of 'cruisie', an oil lamp.*

Raising the Sails

A story, taken from a book by John Mowat, tells of another rather ingenious way in which heather was used by James Bremmner - a famous ship builder and harbour engineer who was born in the Parish of Wick in 1784.

"The brig Isabella of Sunderland was driven on the sands of Dunnet in a storm, and was held fast in the quick sand. Trenches were dug with a view to floating her, but every becoming tide refilled them. Mr Bremmner was a little puzzled and, turning to his foreman said, "John have ye no plan?"

On receiving an answer to the negative, he sharply replied,"Then awa to the hill and poo heather!" Not knowing to what purpose this was meant,

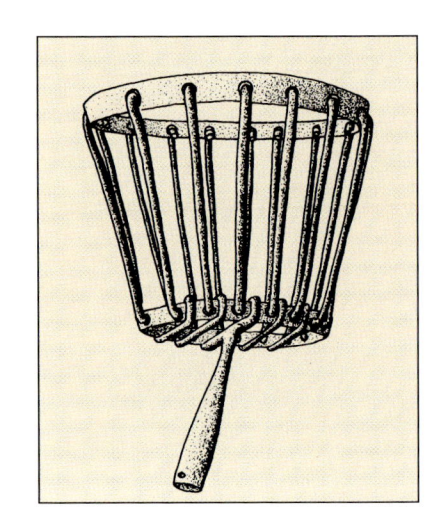

SCOTLAND'S MOST REMARKABLE PLANT

the man quietly submitted and was soon reinforced by a number of women and children from the neighbourhood, organised for the same purpose. On the tide receding, he built up the sides of the trenches with the heather, a plan which effectually prevented them from filling in again. Anchors were put out astern and as the tide flowed, he summoned the whole neighbourhood to pull the vessel off with tackles. The Isabella soon slipped into the water."

From Floor Tiles to Jewellery

Shortly after the second World War, a restriction on the use of wood was initiated. Ground level houses were unable to have the normal timber floor boards and were mainly constructed of concrete or stone. But these floors were hard and cold underfoot, and they had no 'give'. So, in answer to this problem, a small factory employing three to four people, was set up at the side of Loch Lomond, in Dumbartonshire. The factory then set about the production of floor tiles made from the woody heather stem. The resultant tiles were extremely hardwearing and lasted a good length of time.

The tiles were made by compressing the heather stems together into blocks using a special bonding agent. Then they were cut transversely, producing the resultant floor tile.

When eventually restrictions on the use of timber were relaxed, and normal building techniques resumed, production of the heather floor began to dwindle as it proved too expensive to produce.

However the basic technique which had been developed by this small factory of compressing the heather stems was essentially a good one, and was put to a more cost effective use by the jewellery industry.

Initially small blocks were recessed into wood and staghorn to form brooches and pendants. Then a method of dying the stems was developed which resulted in more colourful and interesting jewellery. In time,

Top left: *An original floor tile made from heather stems – extremely hard wearing.*

Top right: *One of the first brooches made from heather stems inlaid in wood.*

Above: *In previous centuries heather was used to dye wool. Today, craft workers and colleges experiment with this natural dye.*

the small craft workshop became more and more sophisticated in techniques, design, production and marketing.

Paint Colourings and Dyes

Born in 1772, Dugald Carmichael, a little-known botanist returned to his native Scotland after a lifetime of exploring the world in search of new plants. Dugald's other great love, besides botany, was painting. But he had great difficulty finding the exact colours he

NINETEEN

INTRODUCING HEATHER

required, so he started to use the natural pigments from plants to colour his paints. And it was to the tops of the heath that he turned to for the colour yellow.

Of course the art of using natural pigments for colouring has been around for centuries with crofters relying on the heather to dye their wool and cloth.

A Traditional Recipe for the Dying of Wool

Gather the tops of the (Barr An Fhraoich) Heather. Gather when they are young and green, and growing in a shady place. Place a layer of wool and heather alternately on the bottom of the pot until the pot is filled. Then add as much water as the pot will hold. Put on the fire to boil, but do not allow to boil dry. The wool will dye a lovely yellow colour which is a good basis for green when indigo is added. If a moss green is require, add gall apples and iron mordant towards the end of dying. Purple and brown tints can be obtained by using old heather tops.

If wanted for winter use, the tips of the heather plant should be picked just before they come into flower. If it is to be used fresh, it can be gathered as long as the flower is in bloom.

The resultant dye is a mordant dye which means the fibre requires special preparation before it can absorb the colour. The treatment is 4oz alum and 2oz cream of tartar to every 1lb of wool.

A Taste of Heather

HEATHER ALE - A Galloway Legend
From the bonny bells of heather, They brewed a drink Lang Syne Was sweeter far than honey, Was stronger far than wine.
R.L. Stevenson

Heather has been used over the years to flavour many different foods and drinks. Little is actually known about the early beverages of Scotland. However, many tales are told of brewing ales and wines from heather flowers. One such brew was known as Heather Crap Ale.

TRADITIONAL RECIPE FOR HEATHER ALE
Ingredients: Heather, hops, barm, syrup, ginger and water.
'Crop the heather when it is in full bloom, enough to fill a large pot. Cover the croppings with water and set to boil for one hour. Then strain into a clean tub. Measure the liquid and for every dozen bottles add one ounce of ground ginger, half an ounce of hops and one pound of golden syrup. Bring to the boil again and simmer for 20 minutes. Strain into a clean cask. Let it stand until milk-warm and then add a teacupful of good barm. Cover with a coarse cloth and let it stand till next day. Skim carefully and pour the liquid gently into a clean tub so that the barm is left at the bottom of the cask. Bottle and cork tightly. The ale will be ready for use in 2 or 3 days and makes a very refreshing and wholesome drink as there is a good deal of spirit in heather.'

As recently as 1993, an Alloa brewery went into production of Heather Ale using an ancient recipe.

TRADITIONAL RECIPE FOR HEATHER WINE.
1¼lbs. Heather Tips (in full bloom)
1 Gallon water
3-4 lbs. Sugar (according to sweetness desired)
2 Lemons
2 Oranges
1 teasp. dried yeast
1 teasp. yeast nutrient.

Cover heather with the water and boil for one hour. Strain off liquid and measure. Restore to one gallon, and add sugar. Stir until completely dissolved. When the temperature drops to 70F, add yeast and nutrient. Leave for 14 days. Then strain into fermentation jar, and when fermentation ceases, strain and bottle. Keep for at least six months!

HEATHER TEA
Gather the flowering heather and after breaking off the hard

SCOTLAND'S MOST REMARKABLE PLANT

woody pieces, spread it in a cool open space and leave for approximately 12 - 16 hours. This should, in theory, allow a slight wither to take place - but with heather having a hard leaf, this is not too noticeable.

Put the heather into a liquidiser and bruise and break-up the heather as much as possible. After this spread thinly in a cool place and leave for a minimum of 3 hours to allow a ferment to take place. This should be apparent from a darkening of the mash. After this, put into an oven, temperature 200-250F until the heather is dry and crisp. The tea retains its misty mauve colour and looks attractive. Used on its own, the product gives a thin liquor. Mixed in equal parts with ordinary tea however, it gives a much stronger flavoursome brew. This is a proper tea - not herbs masquerading as tea.

Tinkers Tea

Trout fishermen having a day on the loch use the following method to make tea – they fill the kettle with loch water and take it to the shore, a sprig of heather and tea is then deposited in the kettle. Next, set old dry heather under and make a mountain of heather over the kettle and ignite. By the time it has burnt out the tea is ready and has a heathery flavour. This method was described to me by Mr George Sproat, 4 Rockfield Road, Tobermory, Isle of Mull.

On the Isle of Skye they had a very simple remedy for tea which had been ruined by smoke from the fire. The solution - a sprig of heather simply placed in the cup!

Heather Whisky

It is said that some of the finest brands of whisky derive some of their most delicate flavours from the heather.

At the Highland Park Distillery, in Kirkwall, Orkney, there was a peculiarly shaped timber building, referred to as the 'Heather House'. This was where heather, which had been gathered in the month of July when the plant was in full bloom, was stored. Carefully cut off near the root, and tied into small faggots of about a dozen branches each, the heather was used on the peat fire to help dry the malt and impart a delicate flavour which, was claimed, to give Highland Park Distillery its unique taste.

It is interesting to note that in former times the wooden containers for fermentation, known in whisky distilleries as 'washbacks', would be cleaned using heather besoms. And when new stills were installed, bundles of heather would be placed in the water and boiled in order to

A selection of products made from heather including wine and honey.

Cliff Jones

sweeten the still before the first distillation took place.

In the nineteenth century and possibly even earlier, illicit stills were used to make whisky - in broad daylight. The crofters were able to do this because, by gathering up and using old stumps of burnt heather, they could make a fire without smoke, and so not raise suspicion!

Heather Honey

There is no other honey quite like heather honey.

Quite different even physically from all other honeys, pure heather honey is sought after by the epicure and commands a high price. Bright golden brown with a pronounced and characteristic flavour, the harvest of heather honey is the premier honey crop in this country.

In some respects, gathering the honey from heather is easier than gathering honey from any other flower source. There is little likelihood of bees swarming when taken to the heather, routine inspection of the hives can be dispensed with and the expectation of a good harvest is reasonably certain - dependent on good weather and no early frosts.

Transportation of the hives to the heather moors is generally undertaken between the end of July and the 12th of August, although this can vary according to the season. However it is advisable to try to catch the best of the Bell Heather and Cross Leaved Heath Crops when they are in the first flush of bloom. Transportation of the bees is best carried out either in the cool of the evening or the early hours of morning. This reduces losses by suffocation.

Due to the flowering structure of the heather plants, where there are numerous flowers on spikes, close to one another in vast expanses of bloom, a considerable amount of honey can be collected in a comparatively short time. Being bell-shaped, the flower is easily entered with the nectar readily available to the visiting bee. The corolla tubes of these small flowers are approximately 2-3mm long with the nectar being concealed at the flowers base. This is easily sought out and collected by the honey bee's spoon-tipped tongue which is approximately 6mm long. The nectar is converted to honey by the bees themselves.

Bell Heather honey is a thinner honey with a port wine colour and a strong characteristic flavour, whilst Cross Leaved Heath Honey is much thinner and lighter in colour.

Weather Predictions

Even predictions in the weather have been associated with heather. It is said, in Scotland, that an extremely rich blossom on the heather during August and September, is followed by severe weather in winter. Whilst another widely held belief, particularly throughout the south of the country and the Cheviot range, is that the burning of the heather 'doth draw doon the rain'!

Plant Badges of the Clans

As already mentioned, Heather can be used, in conjunction with deerhorn, silver and pewter, to make colourful and effective jewellery. But another important decorative use for heather was as a plant badge of the clans. This was used long before the tradition of heraldic badges with the appropriate chief's crest, straps, buckles and mottos.

Referred to as 'Heather Taps', these natural plant badges were worn by the Highlanders in the seventeenth century, if not before, and were placed behind the crest in the bonnet. Heather (Fraoch) was the emblem of the clans MacAlister, MacDonell, Shaw, Farquharson, MacIntyre and Mac Donald, with white heather (Fraoch Geal) pertaining to MacPherson.

It is said that the chiefs of the clan Donald carried into battle, as an emblem of their race, a bunch of wild heather hung from the point of a quivering spear.

Another way in which heather was used decoratively was in the form of dirk handles. Made from the stems and roots

SCOTLAND'S MOST REMARKABLE PLANT

of the plant and carved deeply in Celtic designs these were worn on kilts around the sixteenth and seventeenth centuries.

The Healing Properties of Heather

The healing properties of heather have been recorded as far back as the middle ages, with books on other herbs and their uses dating even further back to the seventh century.

A German book, written in 1565, describes the famous doctor Paulus Aegineta as using the flowers, leaves and stems to heal all types of sores incuding ulcers - both internally and externally.

Fuchs wrote in 1543 that the healing effect of the plant could ease insect bites. Whilst Matthioulos, who lived round about the same time, used the plant in drug form to heal snake bites, eye infections, infections of the spleen and in preventing the formation of stones in internal organs.

Nicolas Alexandre, a Benedictine monk, wrote that boiling heather stems and drinking the liquid for thirty consecutive days, morning and evening, was sufficient to dissolve kidney stones. He added, that the patient should also bathe in the Heather water.

Heather has even been found to help nursing mothers produce more milk. Schelenz wrote in 1914 that Heather was a household remedy for all sorts of illnesses and complaints. However by the turn of the century, heather, in medical terms was generally associated with the prevention and treatment of stones in the bladder and kidney area.

Since 1930, Heather, referred to by the medical profession as Herba Callunae, has been acknowledged by many doctors and chemists as effective against arthritis, spleen complaints, formation of stones, stomach and back ache, even paralysis and tuberculosis. This remarkable plant, which is quite safe for use by diabetics, is also known to be good for sore throats, gout, catarrh and coughs. Some say it even cleanses the blood getting rid of exzema and fevers.

Medical herbalists, to this day, use Calluna vulgaris in the treatment of certain disorders. Containing tannin and several other components, it is used particularly in the treatment of cystitis (bladder infection), as its action is diuretic and antimicrobial.

In the mountain regions of Europe the plant is still used to make a linement for arthritis and rheumatism by softening the herb in alcohol.

Honey for Hay Fever

Pure Heather Honey is recommended for hay fever sufferers.

Heather was not only associated with curing illnesses - it was, according to the Scots National Dictionary, also used figuratively, to describe ailments and other peculiarities common to country folk. HEATHER ILL This was the description used to descibe constipation of the bowels. HEATHER LAMP A springy step common among people accustomed to walking over heathery ground. The term 'heather lamping', refers to lifting feet high when walking - sometimes called a 'heather step'. Walking with a step high and wide was described as walking with 'heather legs'. HEATHER HEADED Sometimes referred to as 'heather heidit', this is the description given to someone with a rather dishevelled head of hair - and indicates a rustic or country background. HEATHER GOOSE This was the term used to descibe a dolt or ninny. HEATHER PIKER This term was a contemptuous epithet for a person living in a poverty stricken or miserly way. HEATHER WIGHT The name given to a Highlander. HEATHER LOWPER A hill dweller, countryman – known as a Heather-Stopper in Perth. HETHER MAN, HATHER A heather seller. Also found purporting to be a term in free masonry.

INTRODUCING HEATHER

Heathers create a wonderful mosaic of colour.

A Heather Garden

As a plant with so many advantages for the present day gardener, it will come as no surprise to find that Heathers are more popular than ever! Providing colour all year round with foliage and flower, heathers are evergreen and will thrive for many years. Inexpensive to purchase and relatively easy to grow, heathers, once established, will provide a weed free garden which requires the minimum of maintenance.

Easy to propagate, heathers are also relatively free from diseases and pests. Small wonder then, that landscape projects large and small, from industrial sites and motorways to housing developments and car parks, make great use of this plant.

Until recently little scientific work has actually been carried out on the hybridisation of heathers, but yet, hundreds of different cultivars are now available from specialist nurseries. Many of these cultivars have been found in the wild as 'chance' seedlings, perpetuated by vegetative propagation. Other cultivars arrive as 'chance' seedlings in nurseries - as sports or mutations in gardens. Each with their own distinctive qualities.

For instance, plants brought back from the remote island group of St. Kilda, (approximately 50 miles west of the Outer Hebrides and 100 miles from the mainland), are extremely dwarf and spreading.

They remain so even in cultivation, their characteristics having been developed to cope with the extreme exposure experienced on the islands.

Cultivars discovered quite by chance include one which was found growing as a sprig on Calluna 'County Wicklow'. 'County Wicklow' has double pink flowers, but this cultivar had double white flowers. The sprig, found in a garden in Argyll, has since been propagated successfully and is now catalogued under the name Kinlochruel. (A selection of cultivars are listed towards the end of this section.)

The heather gardener has many cultivars to choose from and one glance at the photographs in this section should help to convince even the most reluctant gardener that, together with a few conifers and shrubs, an attractive and easily maintained garden can be easily achieved.

Planning

With such a vast range of cultivars to choose from, it would be criminal if we didn't use them to their best advantage in the landscape. A wonderful mosaic of colour can be achieved simply by giving sufficient thought at the planning stage. This is a rough guide which will, hopefully, help you achieve the most from your heather garden. Remember, it is always best to resist the temptation to plant before you have duly considered the site, colour and contasts you want to achieve. We all have a tendency to be impatient and want our plants established - even before they've been planted!

From experience, it is well worth spending the long winter evenings with pen and paper, designing and preparing planting plans. I always think this is half the fun of a heather garden anyway - imagining how they are going to look in a few years. So remember, always plan on paper before buying and planting.

More often than not, we see unimaginative landscape layouts. Why is it that people can be so imaginative indoors - with colour contrasts, focal points and features, but tend to leave that imagination indoors! It is equally important to be creative outdoors.

Basic considerations are necessary when planning a heather garden - soil type, compass point, shade, full sun or sloping sites to name but a few.

The first step however to planning is to survey your site. Measure the areas to be designed, taking particular note of the aspect, the light factor, existing trees and eyesores to be screened. Once this has been done, make a rough sketch on paper to scale, say one quarter inch to 1 foot or 1:50 in metric. This scale will obviously depend on the size of the area to be landscaped.

Then place a piece of tracing paper over your sketch and work away with a pencil, indicating paths, borders and beds. When you are satisfied that you have a nicely balanced, practical design which is pleasing to the eye from all angles, you're ready for the next stage - planting ideas.

A planting plan can be prepared by tracing over designated border areas on the layout plan and marking each plant with a cross or a dot. Remember, always to allocate each plant enough space to allow for development. In the case of heathers, allow one and a half feet (45cm) between plants as an average planting distance. Generally speaking, a heather garden, no matter how well planned, will require some conifers or shrubs to give that extra dimension - height.

Heathers are best planted in groups, with the size of group, being dependant on the available area. For small areas, I would recommend a minimum of three plants per group - for larger groups, no limit is necessary. More impact is achieved by this method than cramming too many varieties into a given space.

A lovely ground pattern effect can be produced by spacing out groups of foliage varieties, intermittently, throughout the length of the border or bed. With varying shades of green, the effect of this method is best seen during the winter months, when flower is less in evidence. Be careful however, not to place flowers of a similar colour next to one another as this can destroy the effect. Try to achieve interesting colour contrasts and associations by referring to the list of heather varieties available or by visiting established gardens.

I have drawn some sketch plans of complete gardens and specific heather beds to help give you some idea of what, with a little imagination, can be achieved. I have also included a planting plan which gives alternative schemes for spring, summer, autumn and winter colour.

Site and Soil

Heathers prefer an open exposed situation; please keep this in mind when choosing a site.

AVOID dense woodland or under trees, steep slopes.

Plant foliage cultivars (cultivated varieties) in sunny position for the best results.

Planted in a shady position, heathers become "leggy", flowering performance is poor and foliage colour is disappointing.

Ideally, heathers prefer lime free, well drained and cultivated

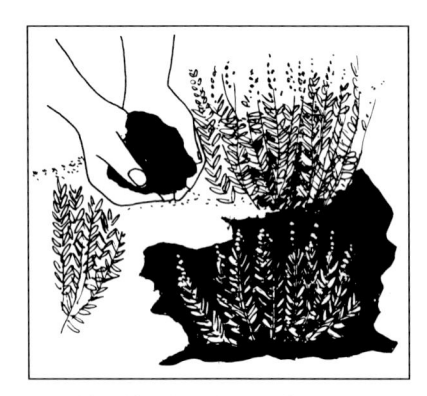

Mulching heathers with peat

top soil. If your soil is on the sandy side, try and build up moisture retaining content by incorporating peat, well weathered composted bark, leaf mould, spent hops or well rotted compost. On heavy clay sites, try to improve drainage by incorporating leaf mould, grit or perlite which will help open up the structure of the clay. Try to provide 20-25cm *8"-10") of workable soil. Remove all weeds, particularly perennial weeds.

Whilst the complete range of heathers will grow happily on an acid soil, (as per rhododendrons and azaleas) don't despair if your soil is alkaline. Certain groups such as Erica carnea, Erica x darleyensis and moderately tolerant Erica vagans, will cope with these conditions. Remember of course that raised beds can be created and suitable topsoil imported to alter the alkaline conditions. Raised beds can be constructed with peat block, timber or stone and need only be approximately 12-18" high.

If in any doubt about your particular soil condition, contact your nearest agricultural college and ask for the horticulturist's advice. A nominal fee may be charged for this service. Alternatively, you could always carry out your own tests with a soil testing kit. Inexpensive and readily available, these kits will help determine the pH value of your soil, indicating its acidity or alkalinity. A pH of 7 is neutral, whilst lower values indicate a more acid soil and higher values - a more alkaline soil. A pH of between 5.5 and 6.4 is considered ideal for all heather groups. Remember that heather plants have very fine roots, therefore the more friable the soil, the faster they will establish in their new positions.

Plants and planting time

The quality of plants from garden centres can vary a great deal. It is obviously better if you can see the plants for yourself before buying. My advice is to be very wary of cheap, mail order plants as these can often be very disappointing, straggly weaklings. It is cheaper in the long run to buy from reputable suppliers, even if it means paying a little more.

What to look for? Well, the

SCOTLAND'S MOST REMARKABLE PLANT

A heather garden approximately four years after planting – (Speyside Heather Centre). Please see layout plan page 30 (bottom right).

first thing to look for is a nicely furnished, bushy young plant, preferably pot grown but not pot bound. Many growers prune back flowering shoots thus ensuring good bushy plants. Avoid straggly, spindly plants which are probably starved. Plants should be labelled, giving time of flowering, colour, eventual height and, obviously, the variety name. One year old plants are usually grown in 8-9cm rigid, plastic pots. Two year old plants are usually sold in 1 or 2 litre pots but are however approximately twice the price.

So as you can see, it can pay to be patient!

If you purchase - and I would recommend that you do, containerised plants, you can plant them, generally speaking, any time. Although care must be taken if planting in summer to ensure that the roots are kept moist until firmly established in the new soil. Obviously however, you should avoid planting in extreme weather when there may be snow, frost or drought conditions. If loose rooted plants are purchased, these should be planted between November and March, when weather conditions permit.

The site for planting should first be thoroughly cleared of any perennial weeds as these are extremely difficult to eradicate on completion of planting, or when the heathers have become more established.

Mix peat well with the top soil prior to planting. The peat should be moist on application. Never plant in pure peat as this will only dry out in summer.

The heather should be planted slightly on the deep side with the lower foliage resting on the ground. If planting in spring, ensure the plants are kept moist throughout the first season as this will ensure that they establish well. A mulch of peat, applied twice a year, after rain, will help to retain moisture, ensure correct conditions and assist in keeping down weeds. Well weathered, pulverised bark can also be used.

Planting distances vary, depending on cultivar and also on personal preference. For a quick ground cover effect, planting distances should be approximately 1ft 6in (45cm) apart - about 6 plants per square yard. Strong growing cultivars can be planted at least 1ft 6in - 2ft (45-60cm) apart with other cultivars (compact or slow growing) approximately 9in - 12in (25-30cm) apart.

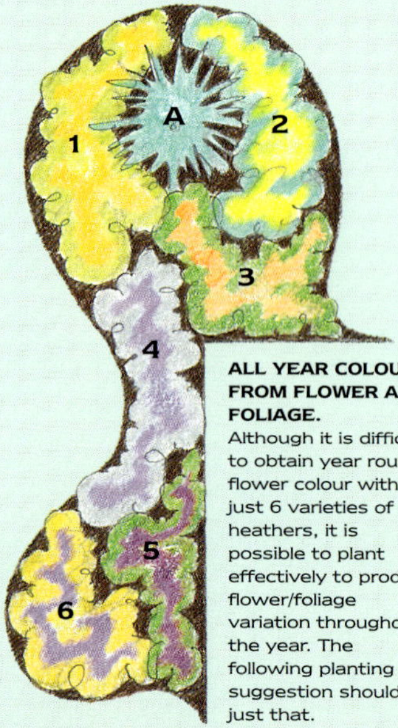

INTRODUCING HEATHER

Suggestions for all 'Scottish connection' heathers

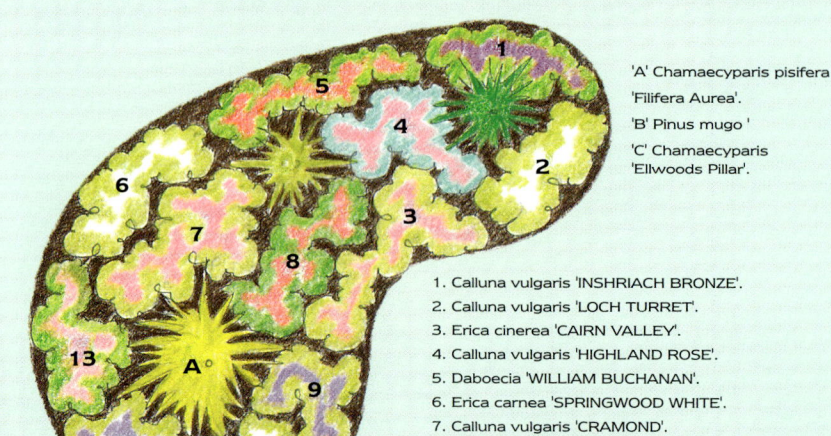

'A' Chamaecyparis pisifera 'Filifera Aurea'.
'B' Pinus mugo'
'C' Chamaecyparis 'Ellwoods Pillar'.

1. Calluna vulgaris 'INSHRIACH BRONZE'.
2. Calluna vulgaris 'LOCH TURRET'.
3. Erica cinerea 'CAIRN VALLEY'.
4. Calluna vulgaris 'HIGHLAND ROSE'.
5. Daboecia 'WILLIAM BUCHANAN'.
6. Erica carnea 'SPRINGWOOD WHITE'.
7. Calluna vulgaris 'CRAMOND'.
8. Calluna vulgaris 'GLEN COE'.
9. Calluna vulgaris 'BOGNIE'.
10. Erica cinerea 'HONEYMOON'.
11. Calluna vulgaris 'HIRTA'.
12. Calluna vulgaris 'SOAY'.
13. Erica carnea 'SPRINGWOOD PINK'.

ALL YEAR COLOUR FROM FLOWER AND FOLIAGE.
Although it is difficult to obtain year round flower colour with just 6 varieties of heathers, it is possible to plant effectively to produce flower/foliage variation throughout the year. The following planting suggestion should do just that.

1. Erica cinerea 'GOLDEN DROP'
2. Erica x praegeri 'IRISH LEMON'
3. Calluna vulgaris 'SPRING TORCH'
4. Calluna vulgaris 'SILVER QUEEN'
5. Erica x darleyensis 'DARLEY DALE'
6. Calluna vulgaris 'ROBERT CHAPMAN'
A. Picea pungens 'Globosa'

THREE TIER CIRCULAR PLANTER – one heather variety per tier

SPRING COLOUR
A. Erica carnea 'CHALLENGER' (Dark red)
B. Erica carnea 'SPRINGWOOD WHITE'
C. Erica carnea 'PIRBRIGHT ROSE'

SUMMER COLOUR
A. Erica cinerea 'STEPHEN DAVIS'
B. Erica cinerea 'GOLDEN DROP'
C. Erica cinerea 'PURPLE BEAUTY'

AUTUMN COLOUR
A. Calluna vulgaris 'DARKNESS'
B. Calluna vulgaris 'SILVER QUEEN'
C. Erica vagans 'MRS F.D. MAXWELL'

WINTER COLOUR
A. Calluna vulgaris 'ROBERT CHAPMAN (Red foliage)
B. Erica x darleyensis 'ARTHUR JOHNSON'
C. Calluna vulgaris 'GOLD HAZE' (Gold foliage)

FOLIAGE COLOUR
A. Calluna vulgaris 'BONFIRE BRILLIANCE' (Flame foliage)
B. Calluna vulgaris 'ROSALIND' (Gold foliage)
C. Calluna vulgaris 'LOCH TURRET' (Fresh green foliage)

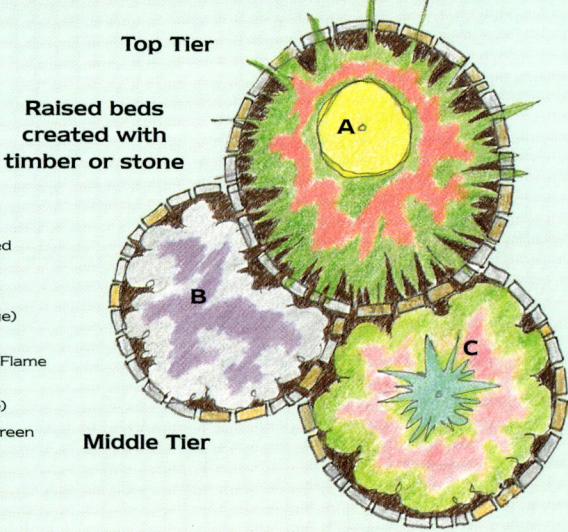

Top Tier

Raised beds created with timber or stone

Middle Tier

Lower Tier

SCOTLAND'S MOST REMARKABLE PLANT

Suggestion for all year round foliage and flower colour

1. Calluna vulgaris 'KINLOCHRUEL'.
2. Erica vagans 'ST KEVERNE'.
3. Calluna vulgaris 'GOLD TURRET'.
4. Erica carnea 'VIVELLII'.
5. Calluna vulgaris 'BEOLEY SILVER'.
6. Erica cinerea 'SHERRY'.
7. Erica carnea 'CECELIA BEALE'.
8. Calluna vulgaris 'DARKNESS'.
9. Calluna vulgaris 'SPRING CREAM'.
10. Erica cinerea 'VELVET NIGHT'.
11. Calluna vulgaris 'BLAZEAWAY'.
12. Erica carnea 'MYRETOUN RUBY'.

'A' Abies procera 'Glauca Prostrata'.
'B' Pinus mugo 'Mops'.

Stepping stones below grass level

Lawn

SUGGESTIONS FOR ALL YEAR ROUND COLOUR - FROM FLOWER AND FOLIAGE.

1. Calluna vulgaris 'ARRAN GOLD'
2. Daboecia 'ALBA'
3. Erica cinerea 'ATRORUBENS'
4. Calluna vulgaris 'SILVER KNIGHT'
5. Erica carnea 'FOXHOLLOW'
6. Erica x darleyensis 'DARLEY DALE'
7. Calluna vulgaris "SPRING CREAM'
A. Chamaecyparis pisifera 'BOULEVARD'
B. Thuja occidentalis 'RHINEGOLD'
C. Chamaecyparis Lawsoniana ' ELLWOODS GOLD'

ALL YEAR ROUND COLOUR FROM FLOWER AND FOLIAGE.

TOP TIER:
1. Erica carnea 'MYRETOUN RUBY'
2. Erica carnea 'SPRINGWOOD WHITE'
3. Erica carnea 'FOXHOLLOW'
4. Erica cinerea 'C.D. EASON'
5. Calluna vulgaris 'BEOLEY GOLD'

MIDDLE TIER:
6. Erica vagans 'MRS D.F. MAXWELL'
7. Calluna vulgaris 'SILVER QUEEN'
8. Calluna vulgaris 'TALISKER'

LOWER TIER:
9. Daboecia 'DAVID MOSS'
10. Erica x darleyensis 'FURZEY'
11. Calluna vulgaris 'BONFIRE BRILLIANCE'
12. Erica tetralix 'ALBA MOLLIS'

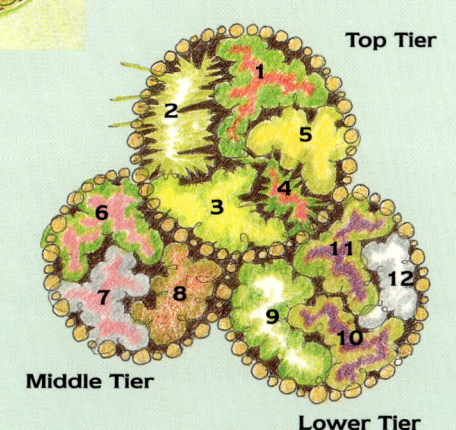

Top Tier
Middle Tier
Lower Tier

Suggestion 'B'
Various heather varieties per tier

INTRODUCING HEATHER

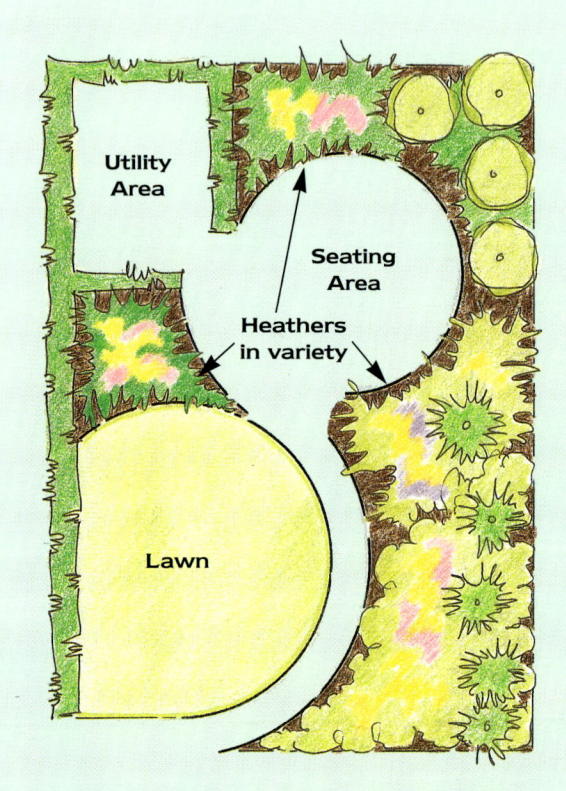

A few ideas for landscaping with heathers.

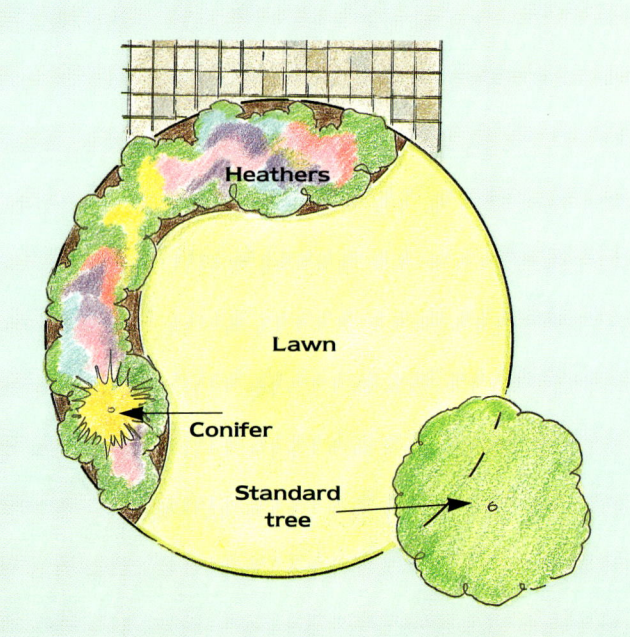

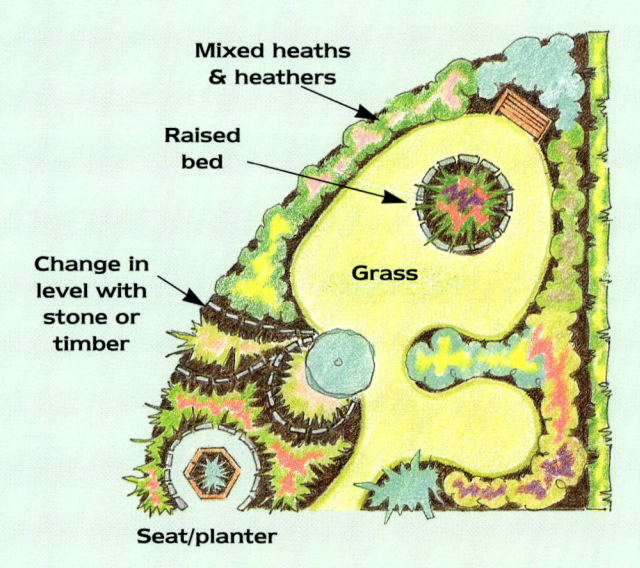

Please see photograph on page 27 for results of plan

Pruning and Propagation

Calluna vulgaris cultivars, I feel, benefit from light annual pruning - removing dead flower heads with secateurs or hedge shears. This will help to keep your plants compact and saves them from becoming straggly. Pruning should be carried out either just after flowering or in early spring, before new growth commences. Remember however that pruning may delay flowering by a week or so. Other species too will benefit from pruning and will help to give nice compact plants with fresh young growths and a better general appearance.

It is always rewarding to increase one's stock by propagating from your own plants. The self-satisfaction at being able to say, "I raised these myself", is quite something.

Propagation of heathers is not too difficult as long as a few basic principles are followed.

Seed

The wild Calluna sets and distributes seeds prolifically. Many 'chance' seedlings are also found in nurseries specialising in the growing of ornamental heathers. The resultant plants however are generally of little garden value.

Layering

The method of layering is a suitable method if only a few plants are required. Here, the lower branches of the plant are pegged down and bent upwards into a prepared mixture of peat and sand. In time, rooting will take place. When a sufficient root system has developed - and I advise you to be patient, sever this rooted layer from the parent plant. Then plant in new permanent position.

Division

Another method, similar to layering except this time the older plants may be planted with only their tips showing. In time these root and are then divided for planting. I would not, however, recommend this method.

A section of the garden at Speyside Heather Centre, showing Silver Birch underplanted with spring flowering 'heathers' – Erica carnea 'Myretoun Ruby' in foreground backed by Erica carnea 'Springwood white' and Erica carnea 'Pink Spangles' on the right.

Cuttings

By far the best method for increasing your stock is to take cuttings. And for the best chance of success it is advisable to cut from plants less than five years old.

Different groups of heather are propagated at different times, but basically it is simply a matter of judging the best time - when growths are not too hard or too soft. Generally speaking, July - August is a good time.

Cuttings should be taken approximately 1 inch in length of non-flowering young shoots. Insert these into a mixture of

INTRODUCING HEATHER

Below: *An old tree stump creates a nice natural feature, the root 'webbing' is interplanted with heathers and alpines.*

Silver birch and heather in association at Tulloch Moor, Strathspey.

Left: *Underplanting silver birch with heathers at Speyside Heather Centre.*

Natures own 'planning & planting' heather and bluebells around an old tree stump in Speyside.

THIRTY-TWO

half peat and half sharp sand - plastic pots and trays can be used. Lightly firm and water using a fine rose on the watering can - keep the cuttings moist and shaded, and spray periodically, making sure that they are shielded from hot sun. I'm sure you'll be impressed by the success rate.

Place in a garden frame or greenhouse, ventilate to prevent fungal rot. When well rooted pot into 8cm pots using acid or ericaceous compost. Never be frightened to nip back any long growths as this will ultimately help to produce a nice bushy plant.

Overwinter in well ventilated garden frame and plant in the garden when one year old.

Pests and Diseases

Generally speaking, pests refers to rabbits, moles and roe deer and not to parasitic pests you might normally associate with garden plants.

In certain areas, rabbits can cause havoc and very often the only positive solution is to erect small mesh wire netting fences around the beds. Mature plants can, however, withstand a little grazing without being too seriously effected.

Moles too can cause damage with destructive burrowing causing damage to the root system of the plant, leaving it suspended in space. This can in turn expose the root sytem and cause it to dry out.

The disease most commonly associated with heather is Browning or Erica Wilt (Phytophthora cinnamomi).

Wet soil conditions can encourage this soil borne disease, which can prove extremely serious in commercial nurseries as it can often render a whole crop unsaleable. The symptoms are flagging and silvering of the leaves which eventually wither and die.

In small gardens, since it is a soil borne disease, any effected plants should be lifted and burned - and the soil in which they were growing, lifted and carried away. New soil should then be put on the garden. It should however be pointed out that attacks of this kind in private gardens are extremely rare.

Another way in which heather can be damaged is with Snow Mould. This is caused as a result of snow lying on the heather for a long period of time. When the thaw arrives, greyish white cobweb-like fungus appears on green shoots. It should be noted however that this is only really common on high moors.

Damage caused by Weather

Winter browning or frosting which is caused by the cumulative effects of bright sunshine, frost and drying winds can damage large areas of heather during the winter months. It is caused when the heather plant loses water to the dry air faster than the roots can replace it from the cold soil. This results in leaves which turn brown and die.

Fertilizers and Weed Control

Heathers require little in the way of fertilisers. When planting, a little bone meal can be applied, dusted around the plant and watered thoroughly or alternatively mixed with the peat in the initial soil preparations prior to planting.

Liquid seaweed based fertilisers (such as Maxicrop) can be applied in late spring to give the plants a boost. On alkaline soil, if you wish to grow the acid loving group of heathers, treating with IRON SEQUESTRENE will be of great benefit.

PHOSTROGEN is also perfectly safe.

Apply all fertilisers at the quantities prescribed on the packet.

If the ground is carefully prepared as described prior to planting, weed control should not be too difficult a task. And after a few years the heathers will spread and cover the ground, making it difficult for weed seeds to penetrate and germinate.

Casaron G is a weedkiller which can be used up to 2 years after heathers have been

INTRODUCING HEATHER

Calluna vulgaris 'Rosalind' – a golden foliage cultivar which has attractive pink flecks. Pink flowers in August/September.

planted. Applied in granular form at the rate of 2oz (mixed together with the same volume of sand) for every 4 square metres. This will keep the area clear for approximately one season. Casaron G should be applied in spring.

Tumbleweed, in jelly form, can be applied to leaves and stems of any perennial weeds which have managed to penetrate the heather plants.

Containers and Window Boxes

Compost: A typical compost suitable for planting heathers in window boxes and containers is made up as follows. By bulk, 2 parts soil, 3 parts peat and 2 parts coarse sand or crushed rock.

A seaweed based fertilizer such as Maxicrop can be used quite safely, right throughout the growing season, if you feel the plants require a boost.

Heather as Dwarf Evergreen Hedges

On occasions I have seen varieties of Erica x darleyensis planted as dwarf hedges around rose beds. These forms are vigorous enough to serve this purpose. A light annual trim is all that is required to keep this dwarf, flowering evergreen tidy. Other types which can also be used in this way are Erica arborea Alpina and Erica terminalis.

Shrubs suitable for the Heather Garden

Many shrubs may be used very effectively in the Heather Garden, providing further flower and contrast of foliage colour. These should be sited very carefully to add height and form, they may be planted as a single specimen or as groupings.

Berberis 'Rose Glow'. One of the many purple leaved Berberis, but different in that it has pink streaks through the purple leaves – a lovely plant which contrasts well with Calluna 'Silver Queen' or Calluna 'Beoley Gold' etc. – approx. height 1m-1.5m. *Berberis 'Harlequin'.* Similar to above but pink streaks are more evident. *Pontentilla 'Longacre'.* Another very fine plant – low spreading habit; good ground cover – yellow flowers. Season June-October.

There are a great number of Potentilla varieties. All of them are, to my mind, suitable for the heather garden. They are extremely hardy. (They are planted above the Car Park at the Cairngorm Chairlift at an altitude of 2,200ft – and thriving), and have a very long season of flowering.

Hebe pagei A splendid ground cover plant, grey foliage which can contrast well with other plants, and white flowers in May-June. 30cm high: 1-1.5m spread. *Acer palmatum. 'dissectum' Atropurpureum.* Leaves finely cut, purple,

An attractive 'island' bed of mixed heathers and dwarf conifers. Some suitable conifers include – Pinus mugo pumilio, Pinus strobus 'Nana', Picea pungens 'Globosa', Chamaecyparis Lawsoniana 'Ellwood's Gold' and Juniperus in variety.

approx. 1m. *Rhododendrons.* can also be included in the heather garden but one should be careful when choosing and siting as they can be overpowering and may detract from the heathers. *Pernettya mucronata.* Approx. 1m. Evergreen, small, dark, glossy, spiny tipped leaves. Male and female forms. At least one male plant needed to three female forms to produce beautiful white, mauve, pink and red berries in Autumn. *Genista hispanica.* Spanish Gorse; a dense cushion plant with spiny leaves. Golden flowers in May/June. *Cytisus Praecox (Warminster Broom).* A lovely cascading broom, cream coloured flowers in abundance in May. 1-1.5m. *Acer palmatum. (Japanese Maple).* Require a more sheltered position but the reward when this shrub matures is worth any effort to cultivate it properly. Beautiful rounded shrub, foliage turning red, orange and yellow in Autumn. A 'gem' – eventually 2m plus. *Acer palmatum. 'Atropurpureum'.* Habit similar to previous, not as vigorous, palmate leaves, lovely purple, in sunshine red purple. A beautiful plant.

Many conifers are suitable for the Heather Garden, but space does not permit other than a mention.

INTRODUCING HEATHER

The beautiful bright golden foliage of Erica arborea 'Albert's Gold' – bright even on the dullest of days.

Tree Heathers

No book on heather would be complete without at least mentioning tree heathers, although, here in Speyside, they are not too successful. I have tried planting Erica arborea Alpina forms without much success, the climate here cutting them right back in winter. Even after eight years from planting, these plants are still only about half a metre in height.

Protection in the form of dead bracken leaves encased in a double layer of netting wire and draped over the plants in tent form, can be used as a protective cover during the winter. Alternatively you could try using branches of Spruce or Pine.

In milder areas however, these plants are extremely useful as a means of adding height, form and flowering to the Heather Garden, between 'seasons'. The varieties you might like to try for this purpose are;
Erica arborea 'Albert's Gold' - Height to 200cms White scented flowers – May. Lovely bright yellow foliage – has proved to be reasonably hardy here in Speyside.
Erica arborea 'Alpina' - (The Tree Heath) 150-180cm March-May. White Flowers. Fresh green foliage.
Erica australis (The Spanish Heath) 150cm
April-June. Rosy red flowers.
Erica australis 'Riverslea'
A deeper flowering form than previous. For best result, plant these tree heathers where possible, in a sheltered site and prune back any damaged branches in spring.

SCOTLAND'S MOST REMARKABLE PLANT

Varieties of Calluna vulgaris with a Scottish Connection

ARRAN GOLD: (Named after island in the Firth of Clyde) Aug-Sept. Compact growth. Gold foliage in summer, orange in winter.
Mauve flowers. 35cm.

BEN NEVIS: Aug-Sept. Mid green foliage. White flowers. 30-35cm.

BOGNIE: (named after a farm SE of Forres, Morayshire) Aug-Sept. Mauve flower. Gold foliage turning bronze in winter. 30cm.

BRAEMAR: Aug-Sept. White flowers. 40-45cm.

CRAMOND: (named after a village near Edinburgh). Sept-Oct. Beautiful double pink flower spikes. 35cm.

DICKSONS BLAZES: (named after a former steel works in Glasgow) Aug- Sept. Cream/pink/red foliage in spring, mauve flowers in late summer. 30-35cm.

DRUM-RA: (named after the area of forest near Aviemore, where it was originally found.) Aug-Sept. White flowers. 30-40cm.

GLENCOE: (named after the pass in Argyll). Aug-Oct. Beautiful double pink flowers. 35-45cm.

GLENFIDDICH: Aug-Sept. Mauve flowers. Gold foliage, turning coppery bronze in winter. 30cm.

GLENLIVET: Aug-Sept. Mauve flowers. Gold foliage, turning red in winter. Slow spreading habit. 25cm.

HIGHLAND ROSE: Yellow/gold foliage, turning bronze in winter. Long spikes with rose coloured flowers. 40-45cm.

C.V. 'Kinlochruel' – *a beautiful double flowered white form, flowering in August/September.*

C.V. 'Inshriach Bronze' displaying lovely multi-coloured young growths in Spring.

INSHRIACH BRONZE: (found near Aviemore) Aug-Sept. Mauve flower. Acid yellow/green foliage in spring.
Rich gold foliage in summer turning copper in winter. 30-35cm.

KINLOCHRUEL: (first found in a garden near Colintraive, Argyll.) Aug-Sept. Sparkling white double flowers.
To my mind the most outstanding white heather in cultivation. 30cm.

LOCH TURRET: (named after an area in Perthshire.) June-July. White flower. Lovely fresh green foliage. 30cm.

TALISKER: Aug-Sept. Mauve flowers. Gold foliage in summer. Red foliage in winter. A very compact plant. 25-30cm.

St. Kilda Collection

The Remote, Rugged St. Kilda Islands which lie approximately 100 miles off the West Coast of Scotland, have for centuries been ravaged by wild Atlantic gales and salt spray. One of the relatively few species of plants to survive the onslaught is Calluna vulgaris. By natural selection, completely isolated from any cultivated heathers, the unique Callunas, evolved over many centuries, developed a tight, ground hugging characteristic, which makes them ideal for tubs, rockeries, troughs, etc.

Flower colour is restricted to white and lilac/mauve, but flowering period is quite wide, many commencing flowering earlier than many well known Calluna cultivars. There are, however, one or two good foliage plants, namely 'Hirta' and 'Soay', and other islands and features in the St. Kilda group lend their name to other cultivars, e.g. 'Boreray', 'Mullach Mor', 'Oiseval'.

St. Kilda heathers planted in a softer more domestic environment at the Glasgow Garden Festival.

HIRTA: Mauve flowers. Lemon yellow foliage. 20cm or less.

SOAY: Red foliage in winter. Lavendar flowers in summer. 20cm or less.

BORERAY: Aug-Sept. Compact white flowers. Dark green foliage. 20cm or less.

OISEVAL: July-Aug. White flowers. Light green foliage. Trailing habit. 20cm or less.

MULLACH MOR: Long white flower spikes. Mid green foliage. 20cm or less.

This list is by no means all the heathers with a Scottish connection, only a small selection of the cultivars available.

Heathers – other than Scottish connection varieties

(* Denotes foliage variety - **Calluna vulgaris**.)

ALBA ELATA: Aug-Oct. Bushy habit, free flowering white heather - mid green foliage. 45-60cm.

ALBA PLENA: Aug - Oct. Double flowering, white, tends to revert to single flowers. Old plants, unpruned tend to become floppy. 30cm.

ALBA PRAECOX: June-Aug. White flowers on dark green foliage. 30cm.

THIRTY-SEVEN

INTRODUCING HEATHER

C.V. 'Darkness' displaying beautiful crimson flowers on a compact plant – flowers in August and September.

ALPORTII: Aug-Oct. Free flowering, long crimson spikes. Good as cut flower foliage darkens in winter. 60-90cm.

ALPORTII PRAECOX: July-Sept. Early flowering, red on dark green foliage. 40-45cm.

AUGUST BEAUTY: Aug-Sept. Masses of white flowers on light green foliage. 40-50cm.

BATTLE OF ARNHEM: Oct-Dec. Late, flowering purple. Dark green foliage which darkens in winter. 50-60cm.

BEOLEY CRIMSON: Aug-Oct. Very long, red flower spikes, ideal for cut flowers. 40-50cm.

***BEOLEY GOLD:** Bright gold foliage with white flowers. 30-45cm.
In my opinion, one of the best foliage plants.

***BEOLEY SILVER:** Aug-Sept. Lovely silver foliage with single white flowers. A lovely contrast plant. 30-45cm.

***BLAZEAWAY:** Aug-Sept. Lavender flowers. Fairly vigorous spreading habit. Orange foliage turning fiery red in winter. 45cm.

***BONFIRE BRILLIANCE:** Aug-Sept. Mauve flowers, mainly grown for foliage effect. Bronzy foliage in summer, turning to fiery red in winter. 30-45cm.

CAERKETTON WHITE: June-Aug. The earliest white flowering Calluna. Fresh green foliage tips in spring. 30cm.

CALIFORNIAN MIDGE: Aug-Sept. Compact cushion plant with pink, mauve flowers. 20cm.

COUNTY WICKLOW: Aug-Sept. Lovely double flowering form (shell pink). Mid-green foliage. 25-30cm.

***CRIMSON GLORY:** Aug-Sept. Long crimson-flower spikes. Gold foliage during the summer months turning to orange/flame in winter. 30-35cm.

***CUPREA:** Aug-Sept. Lavender flowers. Copper coloured foliage which deepens to a deep chocolate colour in winter. An old cultivar, but well worth growing. 30cm.

C.W. NIX: Aug-Sept. Crimson flowers on dark foliage. Sparse habit. 45-60cm.

DAINTY BESS: Aug-Sept. Mauve flowers on woolly greyish foliage. A compact plant. 10cm.

DARKNESS: Aug-Sept. Crimson flowers on an upright, compact plant. A lovely plant with dark green foliage. 30cm.

***DARTS GOLD:** Aug-Sept: White flowers on gold foliage. 15-20cm Low growing habit.

***DARK STAR:** Semi-double dark red flowers. Aug-Oct. A sport from 'Darkness'. Neat compact habit - a lovely bright plant. 20cms.

DIRRY: Aug-Oct. Lilac pink flowers with a low spreading habit. 15cm.

ELSIE PURNELL: Sept- Oct. Long flower spikes of pale rose pink (double) flowers. Greyish green foliage - good for cut flower. 50cm.

***FAIRY:** Aug-Sept. Mauve flowers, mainly grown for the lovely foliage colour. Yellow/orange in summer, deepening in winter. 30cm.

***FIREFLY:** Aug-Sept. Crimson flowers. Very upright habit with stiff flower spikes and reddish foliage, deepening in the winter. 45cm.

***FLAMINGO:** Aug-Sept. Mauve flowers. The real attraction is in the pink/red foliage tips in spring. 30cm.

FOXHOLLOW WANDERER: Aug-Sept. Purple flowers. A good ground cover plant with rich green foliage, Vigorous. 30cm.

FOXII NANA: Aug-Sept. Few mauve flowers. A lovely compact 'pin-cushion' like plant. Ideal for a small rockery. 10-15cm.

FOXII FLORIBUNDA: Aug-Sept. Mauve flowers, much more prevalent than Foxii Nana. Neat little plant. 15cm.

***FRED J. CHAPPLE:** Aug-Sept. Lilac/purple flowers. New growth in spring provides cream, pink and red tones. 40-45cm

***GOLDEN CARPET:** Aug-Sept. Mauve flowers. Low spreading habit. Gold foliage in summer, turning to orange red in winter. 15cm.

***GOLDEN FEATHER:** Aug-Sept. Very few mauve flowers. Feathery gold foliage turning to orange in winter. 30-40cm

***GOLD HAZE:** Aug-Oct. White flowers on gold foliage. 45-55cm

HAMMONDII: Aug-Sept. Long spikes, ideal for cut flowers. Upright habit, with dark green foliage. 60cm.

***HAMMONDII AUREIFOLIA:** Aug-Oct. White flowers with new growth in spring - gold. Contrasts nicely with fresh green, older foliage. 45cm.

H.E. BEALE: Sept-Nov. Long spikes of double shell, pink flowers, ideal for floral art. An older variety but always a firm favourite. 45cm.

HIBERNICA : Oct-Nov. Mauve, flowers late - a very useful plant because of this. 30cm.

HIRSUTA TYPICA: Aug-Sept. Mauve/pink flowers which contrast well with the hairy, silver grey foliage. 45cm.

***INEKE:** Aug-Sept. Mauve flowers on rather spindly, yellow foliage. Quite distinctive. Foliage turns to gold in winter. 30-45cm.

J.H. HAMILTON: Aug-Sept. Lovely deep pink double flowers on a low spreading plant with dark green foliage. 20cm.

JOAN SPARKES: Aug-Sept. Pale, mauve double flowers. Feathery dark green foliage. 20cm.

***JOHN F LETTS:** Aug-Sept. Mauve flowers, compact with prostrate habit. Gold foliage in summer, turning to shades of orange and red in winter. 10-15cm.

***JOY VANSTONE:** Aug-Sept. Mauve flowers. Gold foliage in summer turning to orange in winter. In Speyside, the foliage of this variety tends to brown in the winter, leaving the plant rather unsightly. 40cm.

SCOTLAND'S MOST REMARKABLE PLANT

C.V. 'Mrs Pat'. A lovely plant grown solely for the beauty of the delicate pink tips to the foliage.

*KIRBY WHITE: Aug-Oct. White flowers in profusion. Gold tips to new foliage in spring. Vigorous. 40-50cm.

MAIRS VARIETY: Aug-Sept. Long spikes of single white flowers on a very vigorous plant. Ideal for floral art. 60cm.

MARILEEN/MARLEEN: Sept-Oct. Mauve/pink in bud - never open. Unusual effect from this plant. 30-40cm.

MOUSEHOLE: Aug-Sept. Sparse, pale mauve flowers. This plant is grown mainly for its rather mounded habit. Spiky dark green foliage. 15-20cm.

*MRS PAT: Sept-Oct. Mauve/pink flowers are of no consequence. The attraction of this plant is in the lovely pink tips to the foliage - although these can be damaged in winter. 20cm.

*MULTICOLOR: Aug-Oct. Mauve flowers, grown purely for the very flat habit and vivid red foliage in winter. Unfortunately it tends to revert to green foliage which should be cut when seen. 15cm.

MY DREAM: Sept-Oct. Long flower spikes of white tinged pink double flowers. A sport from H.E. Beale. Good for floral art. 45cm.

NANA COMPACTA: Aug-Sept. Mauve flowers, known as the pin cushion heather. Lovely little plant when grown as a specimen. 15cm.

*ORANGE MAX: Aug-Sept. Mauve flowers, spreading habit. Unusual foliage colour - bronze/gold in summer turning to burnished red in winter. 20cm.

*ORANGE QUEEN: Aug-Sept. Mauve flowers. Fairly vigorous plant. Gold foliage in summer turning to orange in winter. 30-40cm.

*OXSHOTT COMMON: Aug-Oct. Mauve flowers on silvery, grey foliage. Vigorous upright habit. 65-70cm.

INTRODUCING HEATHER

PETER SPARKES: Sept-Oct. Long spikes of deep pink, double flowers. Strong growing variety. Flower spikes are ideal for cut flower. 45cm.

RADNOR: Aug-sept. Double shell, pink flowers on a nice compact plant. 25-30cm.

RED FAVORIT: Double crimson flowers - dark green foliage, spreading habit. Aug-Sept. A sport from J.H. Hamilton. 20cms.

RED PIMPERNEL: Crimson flowers. Aug-Nov. Dark green foliage. 20cms.

RED STAR: Double deep pink flowers. Sept-Oct. Deep green foliage, recent introduction - outstanding. 40cms.

*****ROBERT CHAPMAN:** Aug-Sept. Mauve flowers. My favourite foliage plant. Yellowish/orange in summer turning to flame red in winter. A must for every heather garden. 45cm.

*****ROSALIND:** Aug-Sept. Pink flowers on golden foliage plant. Foliage is flecked with pink tips. Upright habit. 35cm.

*****SALMON LEAP:** Aug-Sept. Mauve flowers. Different foliage colour - salmon pink in summer, turning bronze red in winter. 20-25cm.

*****SERLEI AUREA:** Aug-Oct. White, gold foliage all year round. Fairly vigorous. 45cm.

SCHURIGS SENSATION: Sept-Nov. Deep pink, double flowers. A sport from the cultivar H.E. Beale. 45-50cm.

*****SILVER KNIGHT:** Aug-Sept. Mauve, pink flowers on woolly, silver foliage. Fairly erect habit. 30-45cm.

*****SILVER QUEEN:** Aug-Sept. Lavender flowers compliment a woolly, silver foliage. More spready habit than Silver Knight. A first class plant in Speyside. 30-35cm.

*****SILVER ROSE:** Aug-Sept. Rose flowers on silver/grey foliage. Upright habit. A lovely plant. 45cm.

*****SIR JOHN CHARRINGTON:** Aug-Oct. Red flowers. Yellow/gold foliage in summer, turning to fiery red in winter. 40-45cm.

*****SISTER ANNE:** Aug-Oct. Downy grey foliage on a low compact plant. 15cm.

*****SPITFIRE:** Aug-Oct. Pink flowers. Yellow foliage plant with red/pink tinges throughout. Foliage turns orange/red in winter. 30cm.

*****SPRING CREAM:** Aug-Oct.White flowers. New growths of cream on a nice green foliage plant. 35cm.

*****SPRING TORCH:** Aug-Oct. Mauve flowers. Pinky red tips in spring from the new growths, gradually turning to green as season progresses. 45cm.

*****SUMMER ORANGE:** Aug-Sept. Mauve flowers with orange foliage in summer, deepening in colour during winter. 30-45cm.

Above: *The beautiful winter foliage colour of of C.V. 'Salmon Leap'.*

*****SUNRISE:** Aug-Sept. Mauve flowers. Gold foliage in summer, turning orange/red in winter. A compact plant. 30cm.

*****SUNSET:** Aug-Sept. Lilac flowers. Yellow/gold foliage in summer, turning orange/deep red in winter. 15-25cm.

TIB: July-Aug. Double pink flowers. Useful as it is earlier than other double flowering forms. 25-30cm.

*****TRICOLORIFOLIA:** Aug-Oct. Mauve flowers. Mainly grown for the young spring growth which produces a variety of colours including yellow, bronze and red - hence the name. 30-45cm.

WHITE BOUQUET: Aug-Sept. Large double white flowers on a fresh green plant. 25-35cm.

WHITE GOWN: Sept-Oct. White flowers on a vigorous grey/green foliage plant. Good for cut flowers. 60-70cm.

WHITE LAWN: Aug-Sept. White, interesting and, as the name implies very prostrate heather.

*****WICKWAR FLAME:** Aug-Sept. Mauve flowers. Fairly vigorous foliage plant. Gold in summer turning flame red in winter.

*****WINTER CHOCOLATE:** Aug-Sept. Pale mauve flowers. Young cream and pink tips in spring. Foliage green and yellow/orange in summer, turning to chocolate in the winter months. 30cm.

Below: *The appropriately named C.V. 'Spring Torch' provides value for money with a lovely display (as shown) in Spring from new growth – mauve flowers in August/October.*

Cliff Jones

FORTY

SCOTLAND'S MOST REMARKABLE PLANT

One of my favourite plants. C.V. 'Silver Queen' at its best in August/September when the lovely lavender flowers compliment the woolly silver foliage.

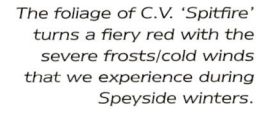

The foliage of C.V. 'Spitfire' turns a fiery red with the severe frosts/cold winds that we experience during Speyside winters.

The beautiful double pink flowers of C.V. 'Cramond' produced in September/October make this an outstanding heather – excellent for floral art.

Erica Cinerea (Bell Heather)

The Scottish hillsides are not only famous for common heather, ling - but they are also enriched with Bell Heather which tolerates the drier slopes and is profusely in bloom throughout late June to September. Fraoch meangan or fraoch a'bhadain in gaelic, the emblem badge of the Clan McAlister. In the Hebrides beer was made from it. This group of plants has also produced many varying forms which are to be found in the wild and in specialist nurseries. I have myself been thrilled to find many variations whilst trekking the hills.

The bell heathers provide the best flower colour range of any groups of hardy heathers. Colours vary from white, shades of pink to bright red and finally beetroot purple. Bi-colour forms are also available.

ALBA MAJOR: July-Oct. White clusters of flowers on lovely fresh light green foliage. 35cm.

ALBA MINOR: June-Oct. White, compact little plant with a long flowering season. Mid green foliage. 20cm.

*APRICOT CHARM: July-Aug. Mauve/pink flowers. Low growing with yellow/green foliage showing tinges of apricot. Colour deepening in winter. 15-20cm.

ATRORUBENS: June-Sept. Ruby red flowers on long sprays. A spreading habit. 30cm.

ATROSANGUINEA: June-Sept. Crimson flowers on dark green foliage. 20-25cm.

CAIRN VALLEY: June-Sept. Lilac pink flowers which tend to fade as the season advances. Dark green foliage. 20cm.

C.D. EASON: June-Sept. Glowing magenta flowers on dark green foliage which turns a lovely dark bronze in winter. 20cm.

INTRODUCING HEATHER

Erica cinerea 'Golden Charm' displays qualities which would make heathers very suitable subjects for hanging baskets – why not?

CEVENNES: June-Sept. Lovely mauve flowers on an upright plant. Prone to damage in severe winters but "worth taking a chance on." 30cm.

C.G. BEST: June-Oct. Salmon pink flowers on tall spikes. This plant has an erect habit. 30cm.

CINDY: July-Oct. Purple flowers on dark green foliage. Upright habit. 30cm.

COCCINEA: June-Sept. Glowing red flowers on very compact little plant. 15cm.

CONTRAST: June-Sept. Dark beetroot coloured flowers on dark green foliage. As the name implies, it is a lovely contrast plant, particularly beside white flowers or yellow foliage. 25-30cm.

DOMINO: July-Oct. White flowers with black sepals and dark flower stalks make this a 'different' white. Nice green foliage. 30cm.

DUNCAN FRASER: July-Aug. Mid green foliage and white flowers tinged with pale pink. 30cm.

EDEN VALLEY: July-Sept. A lovely bi-colour plant the white flowers having lilac tips. Fresh green foliage. 30cm.

***FIDDLERS GOLD:** July-Aug. Purple flowers. Yellow/green foliage tinged pink, deepening colour in winter. 25-30cm.

FOXHOLLOW MAHOGANY: July-Sept. Reddish/mahogany coloured flowers on dark foliage. 25-30cm.

GLENCAIRN: June-Aug. Magenta flowers. Red tips on dark foliage in spring. Compact plant. 20-25cm.

***GOLDEN CHARM:** July-sept. Magenta flowers with yellow/gold foliage. 15cm.

***GOLDEN DROP:** June-Aug. Mauve/pink flowers. Golden yellow foliage which turns to orange and then red as winter approaches. A low spreading habit. 15cm.

***GOLDEN HUE:** June-Aug. Amethyst flowers. Golden foliage in summer deepening in winter. An upright habit. 30cm.

HONEYMOON: July-Aug. White flowers tinged with lavender at the tips. Very compact, low and tight habit. 10-15cm.

HOOKSTONE LAVENDER: July-Sept. Pale lavender flowers on an upright plant. 30cm.

HOOKSTONE WHITE: July-Oct. Long spikes of white flowers on bright green foliage. 35-40cm.

JANET: July-Aug. Pale pink flowers and light green foliage. A compact plant. 25cm.

JOSEPH MURPHY: July-Oct. Purple flowers for a long period. A compact plant. 30cm.

KATINKA: July-Sept. Deep beetroot flowers on dark foliage. Compact, lovely contrast plant. 25-30cm.

KNAP HILL PINK: July-Oct. Magenta flowers on mid green foliage. 30cm.

LILACINA: July-Sept. Pale lilac flowers on upright habit plant with light green foliage. 30cm.

LILAC TIME: July-Sept. Pale lilac flowers on a lovely little plant of tight and bushy habit. Light green foliage. 15cm.

MRS DILL: July-Aug. Magenta flowers on a very compact little plant. 10-15cm.

MY LOVE: July-Sept. Pale mauve flowers on mid green foliage. 20-30cm.

PALLAS: July-Sept. Purple flowers on pale green foliage. Reasonably strong grower. 30cm.

PINK ICE: July-Sept. Deep, clear pink flowers on opening, becoming paler. Mid green foliage with silver reverse. Neat and compact habit. 20cm.

PLUMMERS SEEDLING: July-Oct. Vivid ruby red flowers on a rather straggly plant. But well worth growing for the outstanding flower colour. 30cm.

P.S. PATRICK: July-Sept. Purple flowers on long spikes. Free flowering, upright habit. 40cm.

PURPLE BEAUTY: July-Oct. Amethyst flowers. Large in size with a spreading habit. Long flowering season. Dark green foliage. 20cm.

***ROCK POOL:** July-Aug. Mauve flowers. Bronze/yellow foliage turning red in winter. Spreading habit. 15cm.

ROMILEY: July-Oct. Vivid magenta flowers on a dwarf plant. Lovely. 20cm.

ROSEA: July-Sept. Heliotrope flowers on long spikes. Grey/green foliage. 30cm.

ROSE QUEEN: July-Sept. Magenta flowers. 30-35cm.

RUBY: July-Oct. Beetroot coloured flowers on a bushy compact plant. 30cm.

SHERRY: July-Sept. Blood red flowers on a compact little plant with dark foliage. 25cm.

STEPHEN DAVIS: July-Sept. Glowing magenta flowers on dark green foliage makes this plant outstanding in mid summer. 20cm.

***SUMMER GOLD:** July-Oct. Magenta flowers on golden foliage which makes a pleasing combination. 25cm.

VELVET NIGHT: July-Sept. Dark beetroot flowers make this plant a wonderful foil when placed adjacent to gold foliage. 20cm.

SCOTLAND'S MOST REMARKABLE PLANT

The deep 'beetroot' flowers of Erica cinerea 'Velvet Night' make this an ideal contrast plant – plant alongside gold or silver foliage plant for best effect.

Erica cinerea 'Summer Gold' a very pleasing little plant, the magenta flowers contrasting nicely with the pale green/yellow foliage.

VIOLETTA: July-Aug. Amethyst flowers on a spreading vigorous plant. 25cm.

VIVIENNE PATRICIA: July -Sept. Amethyst flowers which are almost luminous at dusk. 20cm.

***WINDLEBROOKE:** July-Sept. Purple flowers. A vigorous plant with yellow foliage in summer, turning orange/red in winter. 20cm.

* Denotes foliage variety.

Erica Tetralix – the cross leaved heath

Known in Gaelic as Fraoch Frangach, Fraoch an Ruinnse or Fraoch Gucanach, and referred to in English as Ringe Heather - from its use in making scrubbers. Growing in more moist locations, it is this flower which the Clan MacDonald adopted as their emblem badge. Please see photograph on page 46.

***ALBA MOLLIS:** July-Oct. Large white bell flowers on lovely silver/grey foliage makes this a particularly striking plant. 30cm.

ALBA PRAECOX: June-Aug. Earlier flowering plant with white blooms. Grey/green foliage. 30cm.

ARDY: July-Aug. Deep rose pink flowers which are unusual in the tetralix group. 25cm.

CON UNDERWOOD: July-Oct. Deep magenta flowers on grey/green foliage. Long flowering season. 30cm.

HOOKSTONE PINK: July-Oct. Rose pink flowers on grey/green foliage. A lovely delicate colour. 25cm.

PINK STAR: July-Oct. As the name suggests, pink, star-like flowers on soft grey/green foliage. 20cm.

SILVER BELLS: July-Oct. White/flushed pink flowers on a small plant. 15cm.

Although the previous list concludes the three types of heathers indigenous to Scotland, there are other groups of hardy (or relatively hardy) Erica coming under the general term 'Heathers', which make a valuable contribution to the Heather Garden.

Erica ciliaris - DORSET HEATH

Erica carnea - ALPINE or WINTER HEATH

Daboecia cantabrica - ST. DABEOC'S HEATH

Erica vagans - CORNISH HEATH

And two summer flowering hybrid plants;

Erica x praegeri and Erica x watsoni

FORTY-THREE

INTRODUCING HEATHER

The lovely large rose pink flowers (on grey/green foliage) of Erica tetralix 'Hookstone Pink'.

Daboecia cantabrica/scotica

Daboecia azorica (originating from the Azores) and Daboecia cantabrica (originating from Spain, Portugal and the West Coast of Ireland), have hybridised to produce Daboecia scotica. The selection below includes species and hybrids. They each require an acid soil and can easily be damaged by a severe British winter. However they are well worth growing, bearing large bell flowers.

ALBA: July-Oct. Large pure white flowers on light green glossyfoliage. A vigorous plant. 45cm.

BICOLOUR: July-Oct. White, pale pink, purple and striped flowers. A novelty plant with deep green foliage. 35cm.

CINDERELLA: July-Oct. White flowers, flushed pink. Dark green foliage. 45cm.

CORA: July-Oct. Lilac pink flowers on dark green foliage. A small compact plant. 15cm.

DAVID MOSS: July-Oct. White blooms on an upright plant. Dark green, glossy foliage. 30cm.

JACK DRAKE: June-July. Ruby red flowers on a lovely little plant. Foliage is dark, glossy green. 15cm.

POLIFOLIA: June-Oct. Soft purple flowers with greyish/green foliage. 45cm.

PRAEGERAE: July-Sept. Clear salmon pink flowers. Dark green foliage has a purplish sheen whilst the undersides of the leaves are silvery. 45cm

PURPUREA: July-Oct. Bright purple flowers on vigorous plant which has dark green foliage. 45cm.

WM. BUCHANAN: July-Oct. Rosy purple flowers on dark green shiny foliage. 25cm.

Erica carnea

This is commonly known as the Winter or Alpine Heath. The original species was found in the mountain areas of Germany, Austria, Switzerland, Yugoslavia and Northern Italy. Extremely hardy, and flowering in winter or early spring, this species gives good ground cover.

*ANNE SPARKES: Feb-April. Rose pink flowers on yellow foliage which turns bronze/orange in winter. A very slow grower. 15cm.

*AUREA: Feb-May. Lilac-pink flowers on gold foliage which deepens to orange in winter. 20-25cm.

CECELIA BEALE: Jan-April. Early flowering. White on deep green foliage. 15cm.

CHALLENGER: Jan-April. Deep heliotrope flowers on dark bronze foliage. A new variety. 15cm.

DECEMBER RED: Dec-March. Deep lilac-rose flowers. Dark green leaves on a spreading, vigorous plant. 15-20cm.

*FOXHOLLOW: Feb-April. Pale lavender flowers on good foliage. Foliage turns orange in winter months. A low, vigorous, spreading habit. 15cm.

FOXHOLLOW FAIRY: Dec-May. Cream/pink flowers giving a bi-colour effect. Light green foliage and a vigorous, trailing habit. 22-25cm.

HEATHWOOD: Feb-April. Bright rose purple flowers on strong spikes. Foliage dark green, turning bronze in winter. Upright bushy habit. 15cm.

JAMES BACKHOUSE: Feb-May. Rose coloured flowers, mid green foliage. 20-25cm.

KING GEORGE: Dec-March. Deep rose pink flowers freely produced. Foliage is a dark shiny green. A compact growing plant. 20-25cm.

LOUGHRIGG: Jan-April. Rose purple bells in good spikes. Foliage is mid-green until autumn and winter when bronzing occurs. 20cm.

MARCH SEEDLING: Feb-May. Rose pink flowers freely produced on very dark green foliage. Spreading habit and reasonably vigorous. 15-20cm.

MYRETOUN RUBY: Jan-April. Ruby red flowers on very dark green foliage which makes this, a very striking plant. 20-25cm.

PINK SPANGLES: Jan-April. Deep clear pink flowers. Bright green foliage. A lovely plant and a good spreader. 25cm.

PIRBRIGHT ROSE: Jan-March. Rose red flowers. Grey green foliage. Vigorous spreading habit. 15-20cm.

PRAECOX RUBRA: Dec-March. Lilac pink flowers which open early. Deep green foliage. This is a valuable early flowering Carnea. 20-25cm.

RUBY GLOW: Feb-April. Ruby red flowers on dark green foliage. Vigorous spreading habit. 15-20cm.

SPRINGWOOD PINK: Jan-April. Clear pink flowers on mid-green foliage. Vigorous spreading habit. An ideal ground cover plant. 20-25cm.

SPRINGWOOD WHITE: Jan-April. Clear white flowers with chocolate brown anthers. Bright green foliage. An ideal ground cover plant. 20-25cm.

*VIVELLII: Feb-April. Deep red flowers. Dark green foliage, turning bronze in winter. A lovely plant and very popular. 15-20cm.

SCOTLAND'S MOST REMARKABLE PLANT

Erica ciliaris (Dorset Heath)

This is a lovely heath when in full flower and should be grown in an acid soil. As a native of the extreme Southwest Britain, Spain and Portugal, it is not truly hardy in Scotland - although I have known it to survive for a number of years, here in Strathspey.

CORFE CASTLE: Aug-Oct. Salmon pink flowers and mid-green foliage. 20cm.

DAVID MC CLINTOCK: Aug-Oct. Flowers white with pink/mauve tips. Greyish green foliage. 25-30cm.

MRS C.H. GILL: Aug-Oct. Flowers reddish purple on dark green foliage. 25cm.

STOBOROUGH: Aug-Oct. Pearly white flowers on light, fresh green foliage. 30-45cm.

Erica Vagans (the Cornish Heath)

Originating in Cornwall, Northern Spain and Southwest France, this plant will tolerate heavier soils and moderately alkaline soils. It is easily recognised by its closely packed flower spikes.

BIRCH GLOW: Aug-Oct. Vivid rose pink flower. Dome shaped plant with bright green foliage. 20-30cm.

CORNISH CREAM: Aug-Oct. As the name implies, this plant has cream/white flowers on long tapering spikes. Mid-green foliage. 30-40cm.

FIDDLESTONE: Aug-Oct. Cerise flowers freely produced. Fresh green foliage. 30cm.

GEORGE UNDERWOOD: Aug-Oct. Long spikes of pink flowers on light green foliage. 30-35cm.

GRANDIFLORA: Aug-Nov. Shell pink flowers on long spikes. Long flowering season. 30-40cm.

LYONESSE: Aug-Oct. White flowers with gold anthers. Fairly vigorous with fresh green foliage. 35-45cm.

MRS D.F. MAXWELL: Aug-Oct. Cerise/dark red flowers on long spikes. Foliage deep, dark green. This is one of the most popular summer/autumn flowering heaths. 40cm.

NANA: Aug-Sept. Creamy white flowers with brown anthers. Very neat compact plant. 15cm.

Even the snow does not deter Erica carnea 'Pirbright Rose' from flowering.

The lovely large white bells of Daboecia Alba.

Erica vagans 'Mrs D.F. Maxwell' is justifiably one of the most popular summer/autumn flowering heaths.

INTRODUCING HEATHER

ST. KEVERNE: Aug-Nov. Bright rose flowers in profusion. Vivid, fresh green foliage. 20-30cm.

*VALERIE PROUDLEY: Sept-Nov. Very few white flowers. Grown purely for bright yellow foliage. A very compact, slow growing plant. 15-20cm.

WHITE ROCKET: Aug-Oct. Long tapering spikes of white flowers. This is a tall, vigorous plant. 45cm.

Erica x praegeri (summer flowering hybrids)

IRISH LEMON: June-Sept. Lilac pink flowers in clusters . Beautiful young lemon growths in spring. 25cm.

IRISH ORANGE: June-Sept. Lilac pink flowers in clusters. The young growths in spring are yellow flushed with orange. 25cm.

Erica x darleyensis (winter flowering hybrids)

ARTHUR JOHNSON: Nov-May. Lilac pink flowers with light green foliage. a long flowering season. 45cm.

DARLEYDALE: Dec-May. Pale lilac rose flowers on deep green foliage. Vigorous and bushy. 40cm.

FURZEY: Jan-May. Rose purple flowers on strong spikes. Dark green foliage. 40cm.

GEORGE RENDALL: Jan-May. Deep pink flowers. Young pink and yellow growths in spring. 35cm.

J.H. BRUMMAGE: Jan-May. Deep pink flowers. Yellow foliage in summer deepening to golden orange in winter. 40cm.

KRAMERS RED: Magenta flowers. Jan-April. A recent introduction from a breeder in Germany - flower colour similar to 'Myretoun Ruby' - bronze green foliage. Outstanding.

MARGARET PORTER: Jan-May. Clear rose flowers. Shiny green foliage and a spreading habit. 40cm.

SILBERSCHMELZE: Jan-May. White flowers with chocolate anthers. Dark green foliage produces creamy tips in spring. 40cm.

Above right: *A lovely drift of Erica cinerea on the hills above Grantown-on-Spey – with the Cairngorm mountain range in the background.*

Right: *Erica tetralix growing on the more moist slopes, again with the Cairngorms in the background.*

FORTY-SIX

SCOTLAND'S MOST REMARKABLE PLANT

CALENDAR OF HEATHERS AND HEATHS

Flowering time obviously varies according to geographical position, site, aspect, etc. but the following list will give some guide. Remember also that foliage varieties fill any gap in flowering continuity.

JANUARY:

Erica x darleyensis	Silberchmelze
Erica x darleyensis	Arthur Johnson
Erica carnea {	Springwood White
	Springwood Pink
	Vivellii
	December Red
	King George

FEBRUARY:

Erica carnea {	Winter Beauty
	Loughrigg
	Vivellii
	Springwood White and Pink
Erica x darleyensis {	George Rendall
	Furzey
	Arthur Johnson

MARCH:

Erica carnea {	Ruby Glow
	Myretoun Ruby
	Springwood White and Pink
	Winter Beauty
	Vivellii
Erica x darleyensis	George Rendall
Erica x darleyensis	Arthur Johnson

APRIL:

Erica carnea {	Myretoun Ruby
	March Seedling
	Pink Spangles
Erica x darleyensis	Arthur Johnson
Erica x darleyensis	Furzey

MAY:

Erica x darleyensis {	Furzey
	Silberchmelze
	Arthur Johnson
Erica arborea	'Alpina'
Erica erigena	Irish Salmon
Daboecia	Wm. Buchanan

JUNE:

Daboecia {	Wm. Buchanan
	Atropurpurea
Erica cinerea	Praegerae varieties

JULY:

Calluna	Caerketton White
Daboecia	Cultivars
Erica tetralix	Cultivars
Erica vagans	Cultivars

AUGUST:

Calluna vulgaris	Cultivars
Erica cinerea	Cultivars
Erica vagans	Cultivars
Daboecia	Cultivars
Erica tetralix	Cultivars and Summer flowering hybrids

SEPTEMBER:

Calluna vulgaris	Cultivars
Erica cinerea	Cultivars
Erica vagans	Cultivars
Daboecia	Cultivars
Erica tetralix	Cultivars and Summer flowering hybrids

OCTOBER:

Daboecia	Wm. Buchanan
Daboecia	Alba
Calluna {	White Gown
	H. E. Beale
	Silver Rose
	Peter Sparkes
	Elsie Purnell
	My Dream

NOVEMBER:

Calluna {	H. E. Beale
	Peter Sparkes
	Silver Rose
	Underwoodii
	My Dream
Erica vagans	White Rocket and good foliage from Robert Chapman, Blazeaway etc.

DECEMBER:

Erica x darleyensis {	Arthur Johnson
	Furzey
	George Rendall
	Silberchmelze
	a good foliage colour from list of foliage varieties)

FORTY-SEVEN

INTRODUCING HEATHER

The Heather Society

Founded in 1963, to assist in the advancement of horticulture, and in particular, the improvement of, and research into the growing of heaths, heathers and associated plants.

It publishes a Year Book and three Bulletins annually to keep members up to date. It maintains a slide library, provides free technical advice and arranges local and an annual national conference.

For details of membership contact:
Mrs A. Small, Administrator,
Denbeigh, All Saints Road,
Creeting St. Mary, IPSWICH,
SUFFOLK
IP6 8PJ.

Affiliated Societies

North American Heather Society
Secretary: Walter H. Wornick,
Highland View, P.O. Box 101,
Alstead, New Hampshire 03602,
U.S.A.

Nederlandse Heldevereniging 'Ericultura'
Secretary: Mr. J. Dahm,
Esdoornstraat 54, 6681 ZM
Bemmel, Netherlands.

Gesellschaft der Heldefreunde
Chairman: Fritz Kircher,
Tangstedter Landstrasse 276,
2000 Hamburg 62, Germany.

SPEYSIDE HEATHER CENTRE

– was originated in 1972 – by David and Betty Lambie with the intention of being Scotland's Heather Specialists. Located appropriately in the heart of Speyside in the central Highlands of Scotland amidst glorious scenery the award winning centre now attracts approx 85,000 visitors annually.

Any enquiries the reader may have should be addressed to:

Speyside Heather Garden &
Visitor Centre, Dulnain Bridge,
Inverness-shire, Scotland PH26 3PA.
Tel. 047 9851 359
Fax. 047 9851 396

The Author

ACKNOWLEDGEMENTS

Brian Lamb, Medical Herbalist, Castletown, Caithness. Allison McFatridge for helping me obtain some semblance of order to all my research, Sir Kenneth Alexander for forwarding information, Ross Noble, Curator of Highland Folk Museum, Kingussie for information and allowing easy access to his archives, staff at School of Scottish Studies, University of Edinburgh, staff at National Museums of Scotland, Edinburgh, Evelyn Grant of Grantown-on-Spey, Morayshire who initially typed and re-typed!! – Alastair Douglas for the lovely story on page 2, numerous other kind people who supplied information, to all at Nevisprint/The Scottish Collection for their faith, interest, enthusiasm, expertise and professionalism in producing this booklet and finally to my 'big brother' Bob Lambie (who shares my love of natural history) for his photographs, his interest, and for continually encouraging and motivating me to complete this book.